MY REVISION NOTES

Pearson Edexcel

International GCSE (9–1)

GEOGRAPHY

Garrett Nagle and Paul Guinness

T0187406

HODDER EDUCATION
AN HACHETTE UK COMPANY

The Publishers would like to thank the following for permission to reproduce copyright material.

Photo credits

Figure 2.1.3 © Bethany Nagle; **Figure 7.2.3** © P.A. Lawrence, LLC. / Alamy Stock Photo; **Figure 9.1.1** © Rubens Alarcon/Getty Images/Hemera/Thinkstock; all other photos © Garrett Nagle.

Acknowledgements

Figure 3.1.1 Source: https://www.jkgeography.com/tropical-storms---distribution.html; **Figure 3.3.1** Source: https://mrgeogwagg.wordpress.com/2015/09/07/lesson-2-measuring-risk/; **Figure 4.3.1** BP Statistical Review of World Energy 2018, p.10; **Table 4.3.1** data from BP Statistical Review of World Energy 2019; **Figure 4.3.3** BP Statistical Review of World Energy p.50; **Figure 5.2.3** Source: www.indexmundi.com/facts/south-africa/rural-population; **Figure 5.2.4** Source: https://data.worldbank.org/indicator/SP.RUR.TOTL.ZS?locations=ZA; **Figure 6.3.1** Courtesy of Rogers Stirk Harbour + Partners; **Figure 7.1.5** Source: https://upload.wikimedia.org/wikipedia/commons/f/f8/Global_Temperature_Anomaly.svg; **Figure 7.2.2** Source: https://en.wikipedia.org/wiki/Sea_level_rise; **Figure 7.3.1** Source: https://naturalworldheritagesites.org/sites/central-amazonconservation-complex/; **Figure 8.3.1** Source: 2015 Yearbook of Immigration Statistics, Department of Homeland Security, from World Economic Forum; **Figure 9.2.1** Source: https://knoema.com/atlas/maps/GDP-per-capita.

Every effort has been made to trace all copyright holders, but if any have been inadvertently overlooked, the Publishers will be pleased to make the necessary arrangements at the first opportunity.

Although every effort has been made to ensure that website addresses are correct at time of going to press, Hodder Education cannot be held responsible for the content of any website mentioned in this book. It is sometimes possible to find a relocated web page by typing in the address of the home page for a website in the URL window of your browser.

Hachette UK's policy is to use papers that are natural, renewable and recyclable products and made from wood grown in well-managed forests and other controlled sources. The logging and manufacturing processes are expected to conform to the environmental regulations of the country of origin.

Orders: please contact Hachette UK Distribution, Hely Hutchinson Centre, Milton Road, Didcot, Oxfordshire, OX11 7HH. Telephone: +44 (0)1235 827827. Email education@hachette.co.uk Lines are open from 9 a.m. to 5 p.m., Monday to Friday. You can also order through our website: www.hoddereducation.co.uk

ISBN: 9781398321724

© Garrett Nagle, Paul Guinness 2021

First published in 2021 by
Hodder Education,
An Hachette UK Company
Carmelite House
50 Victoria Embankment
London EC4Y 0DZ

www.hoddereducation.co.uk

Impression number   10 9 8 7 6

Year   2025 2024

Cover photo © Vivian Seefeld – stock.adobe.com

Illustrations by Aptara and Barking Dog

Typeset in India by Aptara

Printed in Spain

A catalogue record for this title is available from the British Library.

Schools have a *Licence to Copy* one chapter or 5% for teaching ✓

CCLA Copyright Licensing Agency

Get the most from this book

Everyone has to decide his or her own revision strategy, but it is essential to review your work, learn it and test your understanding. These Revision Notes will help you to do that in a planned way, topic by topic. Use this book as the cornerstone of your revision and don't hesitate to write in it — personalise your notes and check your progress by ticking off each section as you revise.

Tick to track your progress

Use the revision planner on pages 4–6 to plan your revision, topic by topic. Tick each box when you have:
+ revised and understood a topic
+ tested yourself
+ practised the exam questions and gone online to check your answers and complete the quick quizzes

You can also keep track of your revision by ticking off each topic heading in the book. You may find it helpful to add your own notes as you work through each topic.

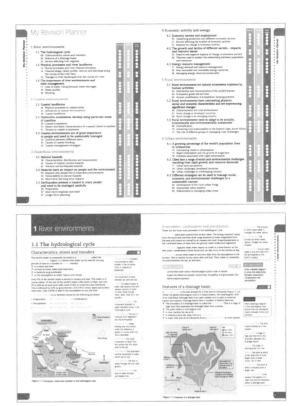

Features to help you succeed

Exam tips

Expert tips are given throughout the book to help you polish your exam technique in order to maximise your chances in the exam.

Now test yourself

These short, knowledge-based questions provide the first step in testing your learning. Answers are at the back of the book.

Definitions and key words

Clear, concise definitions of essential key terms are provided where they first appear.

Key words from the specification are highlighted in bold throughout the book.

Revision activities

These activities will help you to understand each topic in an interactive way.

Exam practice

Practice exam questions are provided for each topic. Use them to consolidate your revision and practise your exam skills.

Summaries

The summaries provide a quick-check bullet list for each topic.

Online

Go online to check your answers to the exam practice and now test yourself questions at **www.hoddereducation.co.uk/myrevisionnotesdownloads**

My Revision Planner

REVISED | TESTED | EXAM READY

Answers to Now Test Yourself and Exam Practice questions at **www.hoddereducation.co.uk/myrevisionnotesdownloads**

REVISED   TESTED   EXAM READY

My Revision Planner

5

7 Fragile environments and climate change

8 Globalisation and migration

9 Development and human welfare

REVISED TESTED EXAM READY

Answers to Now Test Yourself and Exam Practice questions at **www.hoddereducation.co.uk/myrevisionnotesdownloads**

Countdown to my exams

6–8 weeks to go

+ Start by looking at the specification — make sure you know exactly what material you need to revise and the style of the examination. Use the revision planner on pages 4 and 5 to familiarise yourself with the topics.
+ Organise your notes, making sure you have covered everything on the specification. The revision planner will help you to group your notes into topics.
+ Work out a realistic revision plan that will allow you time for relaxation. Set aside days and times for all the subjects that you need to study, and stick to your timetable.
+ Set yourself sensible targets. Break your revision down into focused sessions of around 40 minutes, divided by breaks. These Revision Notes organise the basic facts into short, memorable sections to make revising easier.

REVISED ◯

2–6 weeks to go

+ Read through the relevant sections of this book and refer to the exam tips, summaries, typical mistakes and key terms. Tick off the topics as you feel confident about them. Highlight those topics you find difficult and look at them again in detail.
+ Test your understanding of each topic by working through the 'Now test yourself' questions in the book. Look up the answers at the back of the book.
+ Make a note of any problem areas as you revise, and ask your teacher to go over these in class.
+ Look at past papers. They are one of the best ways to revise and practise your exam skills. Write or prepare planned answers to the exam practice questions provided in this book. Check your answers online and try out the extra quick quizzes at **www.therevisionbutton.co.uk/ myrevisionnotesdownloads**
+ Use the revision activities to try out different revision methods. For example, you can make notes using mind maps, spider diagrams or flash cards.
+ Track your progress using the revision planner and give yourself a reward when you have achieved your target.

REVISED ◯

One week to go

+ Try to fit in at least one more timed practice of an entire past paper and seek feedback from your teacher, comparing your work closely with the mark scheme.
+ Check the revision planner to make sure you haven't missed out any topics. Brush up on any areas of difficulty by talking them over with a friend or getting help from your teacher.
+ Attend any revision classes put on by your teacher. Remember, he or she is an expert at preparing people for examinations.

REVISED

The day before the examination

+ Flick through these Revision Notes for useful reminders, for example the examiners' tips, examiners' summaries, typical mistakes and key terms.
+ Check the time and place of your examination.
+ Make sure you have everything you need — extra pens and pencils, tissues, a watch, bottled water, sweets.
+ Allow some time to relax and have an early night to ensure you are fresh and alert for the examinations.

REVISED

My exams

Paper 1 Physical geography

Date: ....

Time: ....

Location: ....

Paper 2 Human geography

Date: ....

Time: ....

Location: ....

1 River environments

1.1 The hydrological cycle

Characteristics, stores and transfers

REVISED ●

The Earth's water is constantly recycled in a closed system called the hydrological cycle. Figure 1.1.1 shows that water can be held for varying periods of time in a number of stores, namely:

✚ in oceans and seas
✚ on land as rivers, lakes and reservoirs
✚ in bedrock as groundwater
✚ in the atmosphere as water vapour and clouds.

Over 97% of the world's water is stored in oceans and seas. This water is of course saline. Of the rest of the world's water (<3%) which is fresh, just over 2% is held as ice and snow with most of this in Antarctica and Greenland. This is followed by 0.6% as groundwater, and 0.1% in rivers, lakes and surface reservoirs. Only 0.001% is held in the atmosphere at any one time.

Transfers of water occur between stores by the following processes:

• Evaporation
• Transpiration
• Condensation
• Precipitation
• Overland flow
• Infiltration
• Percolation
• Throughflow
• Groundwater flow

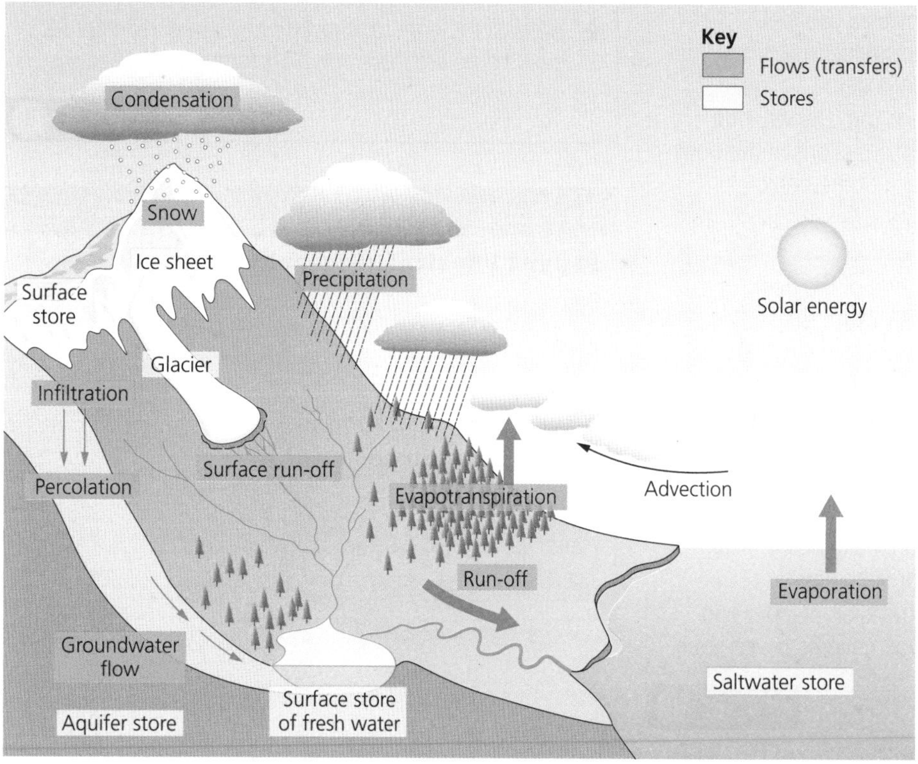

Closed system A system unconnected to other entities. It has no inputs from, or outputs to, elsewhere.

Hydrological cycle The movement of water between air, land and sea.

Stores (of water) Bodies of water that receive, hold and release volumes of water. On land, these include rivers, lakes, reservoirs, and aquifers.

Transfers of water The movement (transfer) of water between stores in the hydrological cycle.

Transpiration The loss of moisture from vegetation into the atmosphere.

Overland flow Water flowing over the surface under the influence of gravity. It occurs when the soil is saturated.

Infiltration The initial movement of water from the surface into the upper level of the soil.

Percolation The downward vertical movement of water within soil or rock.

Throughflow The flow of water through the soil under gravity.

Groundwater flow The flow of water through permeable rock.

Figure 1.1.1 Processes, stores and transfers in the hydrological cycle

Answers to Now Test Yourself and Exam Practice questions at **www.hoddereducation.co.uk/myrevisionnotesdownload**

Evaporation, condensation and precipitation

These are the three main processes in the hydrological cycle.

Evaporation takes place mainly from surface water. The energy required comes from the sun's heat and from wind. Large amounts of water evaporated from the seas and oceans are carried by air masses onto land. Evapotranspiration is the combined losses of water from the ground, water bodies and vegetation.

Condensation happens when water vapour is cooled to a level known as the dew point. Condensation forms clouds and can also occur at the surface as fog.

Precipitation occurs when water in any form falls from the atmosphere to the surface. This is mainly as rain, snow, sleet and hail. Thus, water is constantly recycled between the sea, air and land.

Evaporation The process in which liquid water is changed into water vapour.

Condensation The process by which water vapour changes into water droplets.

Precipitation Occurs when water in any form falls from the atmosphere to the Earth's surface.

Now test yourself
TESTED ◯

1 List the fresh water stores in the hydrological cycle in order of volume.
2 Explain the differences between overland flow, throughflow and groundwater flow.
3 Define evapotranspiration.

Revision activity

Draw a labelled diagram to show the relationship between evaporation, condensation and precipitation.

Features of a drainage basin
REVISED ◯

A drainage basin is the area drained by a river and its tributaries (Figure 1.1.2). While the global hydrological cycle is a closed system, the hydrological cycle of an individual drainage basin is an open system as it is open to external inputs and outputs. Drainage basins have a number of distinct features:

✦ The boundary of a drainage basin is called the watershed. This is a ridge of high land that separates one drainage basin from another.
✦ The point where a river begins is its source.
✦ A river reaches the sea at its mouth.
✦ A tributary joins the main river at a confluence.
✦ A main river and all its tributaries form a channel network or river system.

Exam tip

When drawing a diagram of the hydrological cycle, ensure that you distinguish clearly between stores and transfers.

Drainage basin The area of land drained by a river system.

Watershed A ridge of high land that forms the boundary between two drainage basins.

Source The origin or starting point of a river.

Mouth The point at which a river flows into a much larger body of water – ocean, sea or lake.

Confluence The point at which a tributary joins a larger river.

Channel network (river system) The pattern of a main river and its tributaries within a drainage basin.

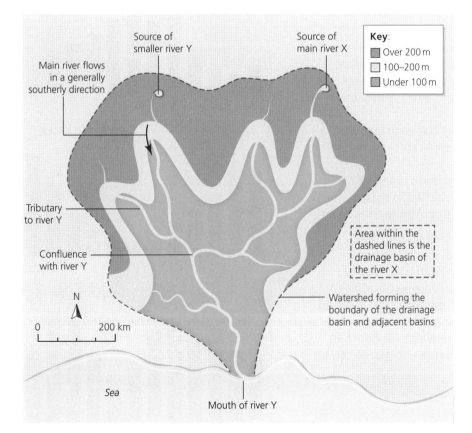

Figure 1.1.2 Features of a drainage basin

The source of a river

The starting point of a river may be:
+ an upland lake
+ a melting glacier
+ a spring in a boggy upland area
+ a spring at the foot of an escarpment.

When small streams begin to flow they act under gravity, following the fastest route down slope. Water is added to them from tributaries, groundwater flow, throughflow and overland flow.

Channel networks

Some main rivers have a large number of tributaries so that no place in the drainage basin is very far from a river. Such an area is said to have a high drainage density. Where a main river has few tributaries, the drainage density is low.

Mouth of a river

Most rivers drain into a sea or ocean, but some drain into lakes which may be far from a coastline. For example, the River Volga, the longest river in Europe, flows into the Caspian Sea.

> **Now test yourself**   TESTED ⬤
>
> 4 Draw a labelled diagram to show the main features of a drainage basin.
> 5 Give two sources of rivers.
> 6 What is a channel network?

Factors affecting river regimes   REVISED ⬤

A **river regime** is the variation in the discharge of a river over the course of a year (Figure 1.1.3). The regime of a river is influenced by several factors:

Climate

The most important factor is climate (Figure 1.1.3):
+ The river Shannon (Ireland) has a typical temperate regime with a clear winter maximum in discharge, the result of high rainfall beginning in late autumn and subsiding in the spring.
+ The river Gloma in arctic Norway has a spring peak associated with snow melt as temperature increases after the cold of winter.
+ The river Po near Venice has two main maxima associated with periods of high rainfall in spring and autumn, and spring snowmelt from Alpine tributaries.

For river regimes the type of precipitation occurring is important, e.g. snow in polar and high altitude regions, and thunderstorms with convection rainfall in warm/hot continental interiors in summer. Temperature has a huge effect on evaporation rates – high temperatures lead to more evaporation with less water getting into rivers. However, as warm air can hold more water vapour than cold air, very high precipitation and river discharge can be experienced in hot, moist climates.

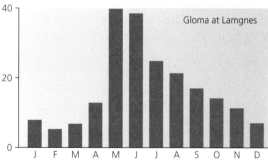

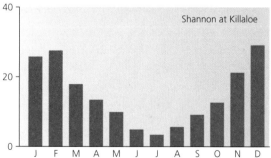

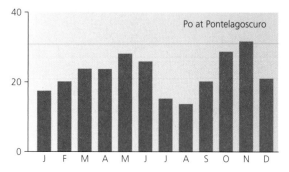

Figure 1.1.3 Three contrasting river regimes for the Shannon, Gloma, and Po

Note: All the graphs show the specific discharge in litres/second per km²

Vegetation

+ High vegetation cover intercepts more rainfall, increases infiltration and reduces overland flow. Broad-leafed vegetation is particularly effective in this respect.
+ In winter deciduous trees lose their leaves and so intercept less rainfall.
+ Wetlands can hold water and release it slowly into rivers.

Geology and soils

+ Permeable rocks allow the accumulation of groundwater which is gradually released into rivers as base flow.
+ The more compact the soil surface, the less infiltration and the greater the overland flow.

Land use

+ Forested areas are most effective in slowing the movement of water to channel networks. In contrast, run off is much faster in areas lacking in vegetation cover.
+ Built environments present the highest levels of impermeability, but urban drainage systems are designed to remove surface water as quickly as possible.

Water abstraction and dams

Water abstraction occurs along most rivers of a significant size. Water is abstracted for human consumption, irrigation and other uses. Abstraction a) directly changes surface water flows and b) indirectly lowers groundwater levels. Dams regulate river flow for the purposes of navigation, irrigation, hydropower production and human water supply. The reservoirs of water held behind dams experience significant evaporation, particularly those in hot, dry climates.

Other factors influencing river regimes

+ Drainage basin size and shape: small drainage basins respond most quickly to rainfall events.
+ Slopes: steeper slopes create more overland flow.
+ Drainage density: basins with a high drainage density respond very quickly to storms.

Storm hydrographs

A storm hydrograph (Figure 1.1.4) shows how the discharge of a river varies over a short time period such as 24 hours and usually refers to a single storm (period of rainfall).

+ Before the storm begins, water is mainly supplied to the river by groundwater flow (base flow).
+ During the storm, some water infiltrates into the local soil while some flows over the surface as overland flow.
+ Overland flow, in particular, reaches the river quickly, causing a rapid rise in the level of the river, the rising limb on Figure 1.1.4.
+ The peak flow is the maximum discharge of the river as a result of the storm.
+ The time lag is the time between the height of the storm and the maximum discharge.
+ The recessional limb shows the speed with which the discharge declines after the peak.

The effect of urban development on hydrographs is to increase peak flow and decrease time lag.

> **Base flow** The normal discharge of a river, which is altered by storm events.
>
> **Water abstraction** The removal of water from water bodies from a surface or underground source.
>
> **Storm hydrograph** A graph showing how the discharge of a river is affected by a storm.

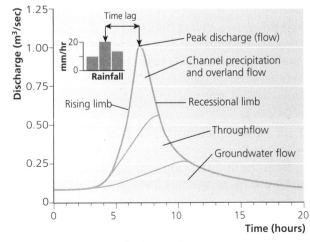

Figure 1.1.4 A storm hydrograph

Exam tip

Remember that a river regime looks at variations in average discharge over the course of a year, while a storm hydrograph records discharge due to a single storm event.

Revision activity

Draw a labelled diagram of a storm hydrograph.

Now test yourself   TESTED ○

7 Define river regime.
8 State two ways in which water abstraction can affect the regime of a river.
9 Explain the time lag on a storm hydrograph.

1.2 Physical processes and river landforms

Fluvial processes and river channel formation   REVISED ○

Rivers have played a major part in forming the landscape in drainage basins through the fluvial processes of erosion, transportation and deposition. However, two other important landscape processes also operate in drainage basins – weathering and mass movement.

Weathering is the breakdown of rock in situ (not involving movement). For example:
+ the freeze–thaw action of physical weathering
+ chemical weathering by rainwater, which is slightly acidic, on rocks.

Biological weathering, particularly the root systems of plants and trees gradually breaking rock apart, is also active in drainage basins.

Mass movement is the large-scale movement of weathered material under the influence of gravity. It carries weathered material into rivers which:
+ contributes additional material to a river's load
+ thereby increasing erosion in the upper course
+ and adding to deposition in the middle and lower courses.

The two main types of mass movement in drainage basins are slumping and soil creep.

Fluvial Of or relating to a river or stream.

Weathering The breakdown of rock in situ.

Mass movement The large-scale movement of weathered material under the influence of gravity.

Friction The resistance encountered when one body moves relative to another body with which it is in contact.

Erosion Wearing away of the Earth's surface by a moving agent such as a river.

Energy and processes   REVISED ○

+ Around 95% of a river's energy is used to overcome friction.
+ The remaining 5% or so is used to erode the river channel and transport this material downstream. The amount of energy in a river is determined by a) the amount of water in the river and b) the speed at which it is flowing.
+ Most friction occurs where the water is in contact with the bed and the banks. Rocks and boulders on the bed increase the amount of friction.

Near the source, rivers channels are shallow and narrow. Also, the beds are often strewn with boulders and very uneven. There is much friction and the water flows more slowly here than further downstream where the channel is a) wider, b) deeper and c) less uneven.

Erosion

There are four processes of erosion:
+ **Hydraulic action**: the sheer force of river water removing loose material from the bed and banks of the river.
+ **Abrasion**: the wearing away of the bed and banks by the river's load.
+ **Attrition**: in swirling water, rocks and stones collide with each other and with the bed and banks. Over time the original sharp edges become smooth and the pieces of rock become smaller in size.
+ **Solution**: some rocks such as limestone dissolve slowly in river water.

Near the source a river cuts down into its bed, deepening the valley. This is **vertical erosion**. In the middle and lower courses, sideways or **lateral erosion** is most important. This widens the valley. Most erosion occurs when discharge is high and rivers are said to be in flood.

Transportation

There are four processes by which a river can transport its load: traction, saltation, suspension and solution. Parts of the load which are moved by traction when the discharge of the river is low may be transported by saltation when the discharge is high.

Deposition

Deposition takes place when a river does not have enough energy to carry its load. This can happen when:
+ the gradient decreases
+ discharge falls during a dry period
+ the current slows down on the inside of a meander
+ the river enters a lake or the sea.

When a river loses energy, the first part of the load to be deposited is the large, heavy material known as the **bedload**. Lighter material is carried further. The gravel, sand and silt deposited is called **alluvium**. This is spread over the floodplain. The solution load is carried out to sea with much of the clay, the lightest suspended particles.

Table 1.2.1 Factors affecting processes: some examples

Factor	Weathering and mass movement	Erosion, transport and deposition
Climate	Chemical and biological weathering most active in hot, wet climates. Cold, dry climates accelerate physical weathering. Slumping aided when weathered material is saturated by heavy rainfall.	Heavy rainfall → higher discharge → increased action of river processes. Higher temperature → increased evaporation → lower discharge and reduced river action.
Slope	Strong relationship between angle of slope and mass movement (increasing shear-stress).	Steep slopes result in fast-flowing rivers with strong erosive power. Gentle slopes encourage deposition.
Geology	More massive rocks tend to be more resistant to weathering than less massive ones. Limestones particularly prone to chemical weathering (solubility).	Rivers erode valleys with soft rock at a rapid rate. Very porous (chalk) and permeable (carboniferous limestone) rocks may lack surface river flow for all or part of the year.
Altitude	Freeze–thaw (physical weathering) very active for long periods in high altitudes.	Snow melt and melting glaciers have a big impact on river regimes and processes.
Aspect	Colder, north-facing slopes in northern hemisphere have higher rate of physical weathering than south-facing slopes.	Higher rates of evaporation and transpiration on south-facing slopes (northern hemisphere) can affect discharge.

Discharge Discharge is defined as the amount of water passing a specific point at a given time (the volume times the velocity). It is measured in cubic metres per second.

Transportation The movement of a river's load by the processes of traction, saltation, suspension and solution.

Load The particles of sediment and dissolved matter carried along by a river.

Deposition The laying down of material carried by rivers or the sea because of a reduction of velocity or discharge.

1 River environments

Revision activity

Draw a labelled diagram to explain the four processes of transportation.

Exam tip

Remember that the factors affecting erosion interact with each other. In any single case, the impact of one factor may be altered through the impact of others.

Now test yourself

TESTED ◯

1 List the three types of weathering.
2 What is mass movement?
3 List the four processes of river erosion.

Channel shape, valley profile, velocity and discharge along the course of the river Tees

The Tees is one of the major rivers in North East England. It drains an area of about 1800 square km. The source of the Tees is on the eastern side of the Pennine mountains. The river flows eastwards to the North Sea. The Tees exhibits most of the classic processes and landforms of the upper, middle and lower courses of rivers. Figure 1.2.1 shows how the long and cross profiles of the Tees change from source to mouth as the river's:

+ gradient decreases
+ depth increases
+ width increases
+ volume increases
+ velocity increases
+ discharge increases
+ sediment size decreases and shape becomes more rounded.

> **Gradient** The degree of slope of the long profile of a river.
>
> **Volume** The amount of water in the river.
>
> **Velocity** The speed of the water.

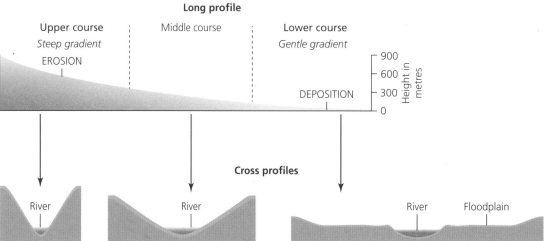

Figure 1.2.1 Long and cross profiles of the river Tees

Upper course

This is mainly an area of moorland where annual precipitation can rise to over 2000 mm per year. The river channel is shallow and narrow. The bed is uneven with sizeable angular boulders in places. There is much friction and the water flows more slowly here than further downstream where the channel is a) wider, b) deeper and c) less uneven.

Vertical erosion has created a steep channel gradient and steep valley sides. Impressive waterfalls are evident along with clear examples of interlocking spurs. High Force is the UK's largest waterfall at 21 metres high. Here a bed of hard rock (dolerite) overlies softer rock (sandstone and shale). As the waterfall has eroded upstream, it has left behind an impressive gorge.

Middle course

Below Middleton-in-Teesdale the valley widens out and the channel slope becomes more gentle. Lateral erosion takes over from vertical erosion, forming distinctive meanders. Good examples can be seen near Barnard Castle. The Tees is joined by important tributaries including the rivers Lune, Balder and Greta. The result is a substantial increase in the volume of water in the river.

Figure 1.2.2 High Force waterfall

Lower course

Here the channel gradient is gentle with the river very close to sea level as it meanders across a fertile clay plain. Deposition is the dominant process. The river has now formed much larger meanders, e.g. near Yarm, across its wide floodplain. Oxbow lakes and levees are clearly evident. The original winding river channel below Stockton has been straightened by artificial cuts to aid navigation. The mouth of the Tees is in the form of a large estuary with mudflats and sandbanks.

Revision activity

Draw a labelled diagram of the long profile of a river valley.

Now test yourself   TESTED ⬤

4   Where is the source of the river Tees?
5   How do the following change along the course of the Tees from source to mouth?
    a)   Gradient
    b)   Depth
    c)   Discharge
6   Name three rivers that join the Tees in its middle course.

Exam tip

The river Tees correlates well with the standard model presented in textbooks. Some rivers do not have such a close relationship for a variety of reasons.

Changes in river landscapes over the course of a river

REVISED ⬤

Upland landforms

The characteristic river landforms in upland areas are a steep V-shaped valley, a steep gradient, interlocking spurs, potholes, and waterfalls, rapids and gorges.

Rivers begin to meander in the upper course. Erosion is concentrated on the outside banks of these small meanders. This eventually causes **interlocking spurs** which alternate on each side of the river. These spurs are ridges of high land which project towards a river at right angles.

Where the bed is very uneven, pebbles carried by fast, swirling water can become temporarily trapped by obstacles in the bed. The swirling currents cause the pebbles to rotate in a circular movement, eroding circular depressions in the bed called **potholes.**

✚ **Waterfalls** occur when there is a sudden change in the course of the river which may be due to differences in rock hardness (Figure 1.2.3).

✚ Waterfalls can form when the hard rock is horizontal, vertical or dipping upstream. The softer rock is eroded more quickly, causing the hard rock to overhang.

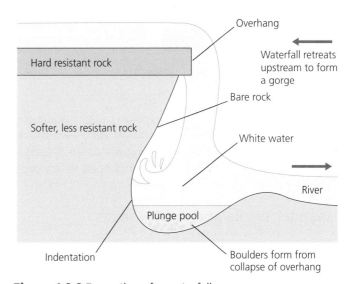

Figure 1.2.3 Formation of a waterfall

✚ The undercutting is caused by abrasion and hydraulic action. The overhang steadily becomes larger until a critical point is reached. When this occurs the overhang collapses.

✚ The rocks that crash down into the plunge pool will be swirled around by the currents. This increases erosion, making the plunge pool deeper.

✚ This process, beginning with the collapse of a layer of hard rock, will be repeated time after time. As a result the waterfall retreats upstream, leaving behind a steep-sided gorge.

Gorge A narrow, steep-sided valley, often formed as a waterfall retreats upstream.

Lowland landforms

Meanders and meander migration

The volume of water increases as more tributaries join the main river. **Lateral erosion** takes over from vertical erosion as the most important process (Figure 1.2.4). As a result, meanders become larger. The current is fastest and most powerful on the outside of the meander, particularly on the downstream section. Erosion is relatively rapid. The outside bank is **undercut**. Again the emphasis is on the downstream section. Eventually it collapses and retreats, causing the meander to spread further across the valley. If the meander has already reached the side of the valley, erosion on the outside bend may create a very steep slope or river cliff. The current on the inside of the meander is much slower. As the river slows it drops some of its load and deposition occurs. This builds up to form a gently sloping slip-off slope (or point bar). Thus, the water is shallow on the inside of the meander and deep on the outside.

> **River cliff** A steep slope forming the outer bank of a meander. It is formed by the undercutting of the river current.
>
> **Slip-off slope** The inside bank of a meander where deposition occurs due to slow river flow.
>
> **Meander neck** A strip of land between two meanders which gradually narrows due to erosion of the outside banks of these meanders.

 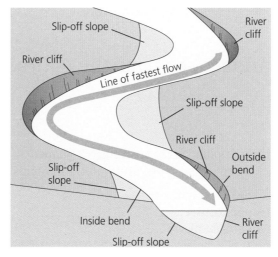 

Figure 1.2.4 Cross-section of a meander

Because of the power of lateral erosion in the middle course, meanders slowly change their shape and position. As they push sideways they widen the valley. But they also move or migrate downstream. This erodes the interlocking spurs, giving a much more open valley compared with the upper course.

Meander necks and oxbow lakes

+ As a river flows towards its mouth, meanders become more pronounced and the valley becomes wider and flatter.
+ As erosion continues to cut into the outside bends of a meander, a meander neck may form (Figure 1.2.5). Eventually, when the river is in flood, it may cut right across the meander neck and shorten its course.
+ For a while water will flow along both the old meander route and along the new straight course. However, because the current will slow down at the entry and exit points of the meander, deposition will occur.
+ After a time the meander will be cut off from the new straight course to leave an oxbow lake.

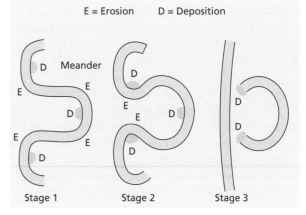

Figure 1.2.5 Formation of an oxbow lake

Floodplains and levees

A **floodplain** is the area of almost flat land on both sides of a river. It is formed by the movement of meanders explained above. When discharge is high the river is able to transport a large amount of material in suspension. At times of exceptionally high discharge, the river will overflow its banks and flood the low-lying land around it. The sudden increase in friction as the river water surges across the floodplain reduces velocity and causes the material carried in suspension to be deposited on the floodplain. The heaviest or coarsest material will be dropped nearest to the river. This can form natural embankments alongside the river called **levees** (Figure 1.2.6). The lightest material will be carried towards the valley sides. Each time a flood occurs, a new layer of **alluvium** will be formed. This gradually builds up the height of the floodplain.

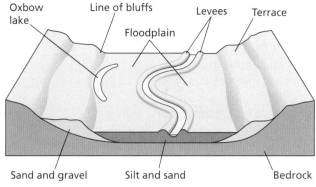

Figure 1.2.6 Cross-section of a river floodplain

Revision activity

Write simple paragraphs to explain the processes operating on the outside and inside banks of a meander.

Now test yourself

TESTED ◯

7 List four characteristics of an upland river valley.

8 Draw a fully labelled diagram of a meander.

9 How are levees formed?

Exam tip

When drawing a diagram showing the formation of an oxbow lake, make sure you label where erosion and deposition has occurred.

1.3 The importance of river environments and their management

Uses of water; rising demand; water shortages

REVISED ◯

Uses of water

Global water use increased from 1.22 trillion m³ in 1950 to over 4 trillion m³ in 2018. This increase was more than twice the rate of population growth. Current global water use by sector is:

+ agriculture: 69%
+ industry: 19%
+ domestic: 12%.

Figure 1.3.1 contrasts water use in developed and developing countries. In developing countries agriculture accounts for over 80% of total water use, with much devoted to irrigation. In developed countries the demand for water for leisure activities has risen significantly.

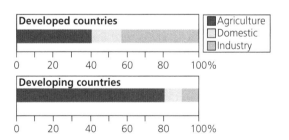

Figure 1.3.1 Water used for agriculture, industry and domestic for the developed and developing worlds

Irrigation Supplying dry land with water by systems of ditches and also by more advanced means.

Rising demand for water

Global water use has more than tripled since the 1950s due to:

+ population growth
+ rising living standards
+ changing patterns of food consumption (from grain-based to protein-based diets)
+ increasing urbanisation
+ higher water demand from industry.

Increasing water supply

The objective in all methods of water supply is to take water from its source to the point of usage. In 2015, about 91% of the global population had access to piped water supply, up from 76% in 1990. However, over 660 million people still do not have access to an improved water source. Much of the increase in water consumption has been made possible by investment in water infrastructure, particularly dams and reservoirs. In some countries water is delivered on a daily basis to urban areas that are not yet connected to the mains supply.

Areas of water shortage

Water shortage is most pronounced in the Middle East and North Africa, Central Asia and northern India.

Much of the precipitation that falls to the Earth's surface cannot be captured and the rest is very unevenly distributed. The arid regions of the world cover 40% of the world's land area, but receive only 2% of global precipitation.

Water scarcity is to do with the availability of potable water. It is threatening to put world food supplies in jeopardy, limit economic and social development, and create serious conflicts between neighbouring drainage basin countries.

+ A country is judged to experience water stress when water supply is below 1700 m^3 per person per year.
+ When water supply falls below 1000 m^3 per person a year, a country faces water scarcity for all or part of the year.

Water depletion hotspots are caused by drought, groundwater depletion, ice-sheet and glacier loss/retreat, surface water loss (the drying of the Aral and Caspian Seas), and the filling of large reservoirs (the Three Gorges Dam).

Water surplus

Water surplus occurs when the demand for water is less than the supply. This situation exists mainly in temperate and tropical wet areas and includes large parts of North America and Western Europe, and the Amazon and Congo Basins.

Countries and regions within countries experiencing water surplus tend to have the following:

+ Positive geographical characteristics with regard to water – high rainfall and surface run-off, large stores of surface water and significant aquifers. Moderate rates of evaporation can also play a major role.
+ Low population density and effective water management (quantity and quality).

Water supply The provision of water by public utilities, commercial organisations or by community endeavours.

Potable water Water that is free from impurities, pollution and bacteria, and is thus safe to drink.

Water stress When water supply is below 1700 cubic metres per person per year.

Water scarcity When water supply falls below 1000 cubic metres per person per year.

Revision activity

List four reasons for the increasing global demand for water.

Exam tip

Remember that water supply is water that can be accessed on a regular basis by those people who want to use it. Investment in infrastructure is usually needed for this to happen.

Now test yourself   TESTED ◯

1 Define potable water.
2 How many people globally do not have access to an improved water source?
3 What is the difference between water stress and water scarcity?

Water quality

Reasons for variations in water quality

Water pollution comes from a number of sources including:

+ contamination by agricultural run-off, particularly from factory farming
+ industrial pollution of rivers and other water bodies
+ urban run-off carrying pollutants from cars, factories and other sources
+ untreated sewage.

Each year, more than 80 per cent of the world's wastewater is released to the environment without being collected or treated. This a) pollutes the environment and b) wastes a renewable resource. While rivers in more affluent countries have become steadily cleaner in recent decades, the reverse has been true in much of the developing world. Rivers in Asia are the most polluted.

> **Pollution** Contamination of the environment.
>
> **Cloud seeding** Releasing particles of dry ice or silver iodide into cold clouds to encourage precipitation.
>
> **Rainwater harvesting** The accumulation and storage of rainwater for reuse on-site rather than allowing it to run off.

The storage and supply of clean water

Figure 1.3.2 shows water supply and management methods in the large Canadian province of Alberta.

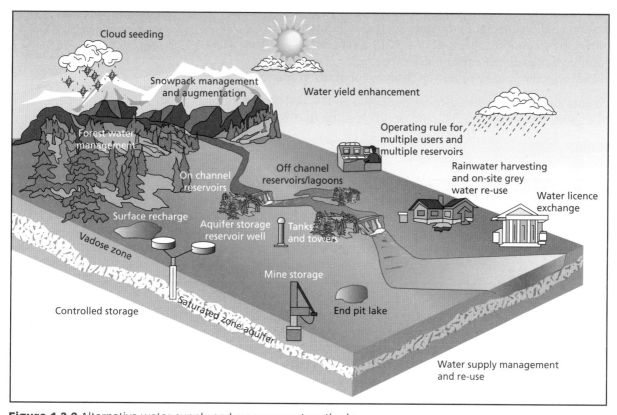

Figure 1.3.2 Alternative water supply and management methods

Dams and reservoirs

+ Used to save, manage and prevent the flow of excess water into specific regions.
+ 'On channel' reservoirs exist where a dam has been built across an existing river (Figure 1.3.2).
+ 'Off channel' reservoirs usually use depressions in the existing landscape or human-dug depressions to store water.

Globally the construction of dams has declined since the 1960s and 1970s because most of the best sites for dams are already in use or such sites are strongly protected. An alternative to building new dams and reservoirs is to increase the capacity of existing reservoirs by extending the height of the dam.

> **Reservoir** A natural or artificial lake used as a source of water supply.
>
> **Dam** A barrier that holds back water.

Wells and boreholes

+ Wells and boreholes are sunk directly down to the water table.
+ Aquifers provide about half of the world's drinking water, 40% of the water used by industry and up to 30% of irrigation water.
+ About 35% of all public water supply in England and Wales comes from groundwater. Groundwater is even more important in arid and semi-arid areas.

Pipelines

In general, water is redistributed through water networks and grids over longer distances in developed countries than in developing countries because of the high cost of water infrastructure. However, water grids are nowhere near as extensive as power grids. Thus, in many countries there is limited ability to move water from areas of water surplus to areas of water deficit.

Water treatment

+ Almost all sources need some degree of water treatment to be pure enough for human consumption. Contamination can be caused by both anthropogenic (caused by human activity) and natural contaminants.
+ Developed countries use central source treatment systems (treatment works). These involve physical processes such as sedimentation and filtration, and disinfectant processes such as chlorination.
+ In developing countries, point-of-use (POU) treatment is also used where treatment works do not exist or lack capacity.
+ Groundwater tends to be relatively clean in comparison to surface water.

Well-maintained urban water systems lose between 10% and 30% of the water they transport. However, in developing world cities, up to 70% can be lost.

> **Now test yourself** TESTED ○
>
> 4 State four sources of water pollution.
> 5 Give two examples of large-scale water transfer by pipeline.
> 6 Define water treatment.

> **Wells and boreholes**
> A means of tapping into aquifers (water-bearing rocks), gaining access to groundwater.
>
> **Aquifer** A rock that allows water to move through it, such as a layer of sandstone.
>
> **Water treatment** Improving water quality to make it acceptable for a particular end-use such as human consumption or irrigation.

> **Revision activity**
>
> How is water supplied to the area in which you live?

> **Exam tip**
>
> The highest level of water treatment is used for human consumption. Water used for industrial and agricultural purposes does not require such a high degree of purity.

Flooding
REVISED ○

Causes of river flooding

+ The primary causes of most floods are external climatic forces. These factors have become more variable and extreme with climate change.
+ Secondary flood causes tend to be drainage basin specific. Over time the impact of human activity on drainage basins has steadily increased.

Rainfall intensity

Floods in the UK are associated with deep depressions (low-pressure systems) in autumn and winter. They tend to be both long in duration and wide in the area covered.

Seasonal variations in discharge

In India, up to 70% of the annual rainfall occurs in one hundred days in the summer southwest monsoon. Melting snow in spring from mountain ranges such as the Himalayas, Rockies and Alps can be responsible for widespread flooding in many countries.

> **Flood** A high flow of water that overtops the bank of a river.
>
> **Monsoon** The season of heavy rain during the summer in hot Asian countries.

Relief

Steep slopes encourage rapid run-off. The potential for damage by floodwaters increases exponentially with velocity and speeds above 3m per second.

Urbanisation

The creation of highly impermeable surfaces increases run-off; a dense network of drains and sewers increases drainage density; natural river channels are often constricted by bridge supports or riverside facilities, reducing their carrying capacity; due to increased storm run-off, many sewerage systems cannot cope with the resulting peak flow.

In addition, impermeable rocks, a high drainage density, lack of vegetation cover and soils with a low infiltration capacity all encourage rapid run-off. Human actions including deforestation and poor agricultural practices can reduce interception and increase run-off.

The prediction and prevention of flooding

Prediction

In recent decades, flood forecasting and warning have become more accurate. This is particularly so in developed countries. According to the US Geological Survey (USGS), flood prediction requires several types of data:

- the amount of rainfall occurring
- the type of storm producing the precipitation
- the rate of change in the discharge of the river/channel network
- the characteristics of the drainage basin.

The task then is to convey information about the immediacy and severity of the flood risk to people likely to be affected as quickly as possible. For example, using the UK government website (www.gov.uk) and entering the name of a location or postcode, you can check if you are:

- at immediate risk of flooding
- at risk of flooding in the next five days
- in an area that's likely to flood in the future.

This website also provides information on (a) how to plan ahead for flooding, (b) what to do in a flood, and (c) how to recover after a flood.

Prevention

- Traditionally floods have been managed by methods of 'hard engineering' such as dams, reservoirs, levees, straightened channels and flood-relief channels.
- Increasingly 'soft engineering' measures have come to the fore. These techniques focus on working with natural processes and features rather than attempts to control them. They include catchment management plans, river restoration and wetland conservation.
- Land-use zoning can reduce the number of premises and people at risk of flooding.
- Hazard-resistant design (flood-proofing) includes any adjustments to buildings and their contents that help reduce losses.

US Geological Survey A United States scientific agency that studies the landscape of the USA, its natural resources, and the natural hazards that threaten it.

Land-use zoning The segregation of land use into different areas for each type of use.

River management in south-western USA

The USA is a huge user of water. The western states of the USA, covering 60% of the land area with 40% of the total population, receive only 25% of the country's annual precipitation. Yet each day the west uses as much water as the east. Water management has been vital to the prosperity of California and other states in this region.

+ Over time there has been a huge investment in water transfer schemes. This has benefited agriculture, industry and settlement.

+ California has benefited most from this investment in water supply. Seventy per cent of run-off originates in the northern one-third of the state but 80% of the demand for water is in the southern two-thirds. Large amounts of water are transferred from rivers in the north to water bodies in the more heavily populated south for irrigation and urban/industrial use.

+ The 2333-km-long Colorado River is an important source of water in the south-west. Over 30 million people in the region depend on water from the river. Major problems over the river's water have arisen because population has increased, along with rising demand from agriculture and industry.

+ Major dams along the Colorado include the Hoover Dam and the Parker Dam.

+ The $4 billion Central Arizona Project (CAP) is the most recent scheme to divert water from this great river. Completed in 1992, it brings vital water supply to the cities of Phoenix and Tucson. A new source of supply was necessary due to heavy depletion of groundwater resources.

+ Resource management strategies include measures to reduce leakage and evaporation losses; recycling more water in industry; charging more realistic prices for irrigation water; extending the use of the most efficient irrigation systems; changing from highly water-dependent crops such as rice and alfalfa to those needing less water.

+ Future options include developing new groundwater resources; investing in more desalination plants; constructing offshore aqueducts that would run under the ocean from the Columbia River in the north-west of the USA to California.

There is now general agreement that planning for the future water supply of the south-west should embrace all practicable options.

7 Describe the imbalance between precipitation and population distribution in California.

8 Why is the Colorado River under so much pressure?

River management in China: The Three Gorges Dam

The Three Gorges Dam across the Yangtze River is the world's largest multi-purpose river management scheme. The dam was completed in 2009 and is over 2 km long and 100 m high. The lake impounded behind it is over 600 km long. The Yangtze basin provides 66% of China's rice and contains 400 million people. The river drains 1.8 million km² and discharges 24,000 m³/second of water annually.

The advantages of the project:

+ Flood control protects 10 million people living downstream from the flooding caused by the seasonal nature of the Yangtze River. Downstream cities include Wuhan, Nanjing and Shanghai.

+ An electricity-generating capacity of 22,500 MW which supplies Shanghai and Chongqing in particular. It is the largest electricity installation in the world.

+ The locks constructed as part of the project allow shipping above the dam. Tourism has benefited from the rapid expansion of cruise ships along the river.

+ Supplementing water supply downstream in dry periods to agriculture, industry and domestic supply.

Many disadvantages have been quoted. These include:

+ Over 1.25 million people were forced to move to make way for the dam and lake.

+ The region is seismically active and landslides are frequent.

+ The dam traps silt which gradually reduces the capacity of the reservoir and reduces the fertility of farmland downstream.

+ The dam interferes with aquatic life.

9 How many people live in the Yangtze basin?

10 What is the energy capacity of the Three Gorges Dam?

Revision activity

List the types of data required for flood prediction.

Exam tip

River management is generally the most important aspect of water management in a country or region, but look back at Figure 1.3.2 to remind yourself of the other aspects of water management.

Exam practice

1 Which is the largest store of fresh water?   (1 mark)
   A Groundwater
   B Ice and snow
   C Rivers, lakes and surface reservoirs
   D The atmosphere

2 a) Define evapotranspiration.   (2 marks)
   b) Compare infiltration and throughflow.   (2 marks)

3 Describe the features of a drainage basin.   (4 marks)

4 Compare the upland and lowland cross-sections of a river valley.   (3 marks)

5 Name one process of transportation.   (1 mark)

6 Explain the formation of a floodplain.   (4 marks)

7 Look at Figure 1.3.2 (water supply and management methods). Discuss four of the methods identified in the diagram.   (8 marks)

Total: 25 marks

Summary

+ The hydrological cycle is a closed system. Water is held in stores, with transfers (flows) of water occurring between stores by a number of processes.
+ The key features of a drainage basin are: source, watershed, channel network, confluence and mouth.
+ The regime of a river is influenced by several factors, including climate, vegetation, geology and soils, land use, and water abstraction and dams.
+ A storm hydrograph shows how the discharge of a river varies over a short period of time.
+ The physical processes of erosion, weathering and mass movement, transportation and deposition give rise to characteristic river landforms.
+ Channel shape, valley profile, gradient, velocity, discharge and sediment size and shape change along the course of a river.

+ Along the course of a river distinctive upland and lowland landforms can be recognised.
+ Human populations use water in a number of different ways. The demand for water has increased significantly over time, requiring greater investment in methods of water supply.
+ Areas of water shortage and water surplus can be identified at both the global and national scales.
+ Variations in water quality occur due to various forms of pollution and the level of investment to supply clean water.
+ The causes of river flooding include rainfall intensity, seasonal variations in discharge, relief and urbanisation.
+ The ability to predict and prevent flooding has improved over time.

2 Coastal environments

2.1 Coastal landforms

Physical processes in coastal areas

REVISED ●

There are a number of processes that occur in coastal zones. These include:
+ wave action from constructive and destructive waves
+ wind action
+ mass movements (sliding and slumping) and weathering (mechanical, chemical and biological).

> **Mass movement** The large-scale movement of weathered material under the influence of gravity.
>
> **Weathering** The breakdown of rock in situ.

Table 2.1.1 Constructive and destructive waves

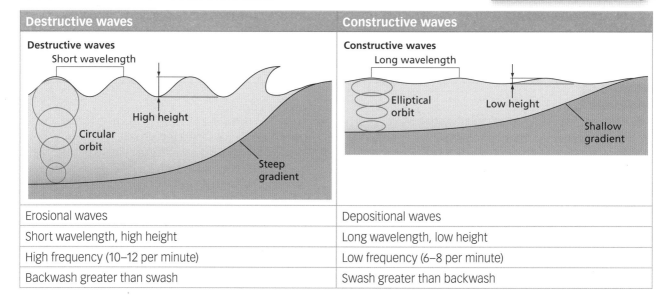

Destructive waves	Constructive waves
Erosional waves	Depositional waves
Short wavelength, high height	Long wavelength, low height
High frequency (10–12 per minute)	Low frequency (6–8 per minute)
Backwash greater than swash	Swash greater than backwash

Erosional processes

Wave erosion includes hydraulic action, abrasion, attrition and solution.

Processes of transportation

In the water, particles are moved in different ways:
+ larger particles are dragged along the sea floor by traction
+ smaller particles may be bounced along the sea floor by saltation
+ very fine materials are held up in suspension
+ dissolved sediments, e.g. calcium, may be carried in solution.

Deposition

Deposition occurs for a variety of reasons:
+ a decrease in wave energy or velocity
+ a large supply of material
+ an irregular, indented coastline, e.g. river mouths.

> **Hydraulic action** The force of air and water when the waves break.
>
> **Abrasion** The wearing away of cliffs by the load carried by the sea.
>
> **Attrition** The wearing away of the load carried by the sea.
>
> **Solution** The removal of chemical ions, especially calcium, which causes rocks to dissolve.

Wave refraction and longshore drift

Wave refraction occurs when waves approach an irregular coastline or at an oblique angle. Refraction reduces wave velocity and, if complete, causes wave fronts to break parallel to the shore. Wave refraction concentrates energy on the flanks of headlands and disperses energy in bays. However, refraction is rarely complete and consequently **longshore** or **littoral drift** occurs. The **swash** is the movement up the beach while the **backwash** is the movement down the beach.

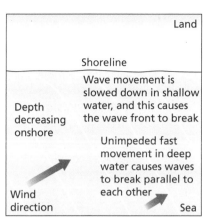
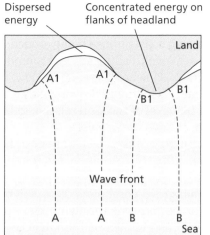
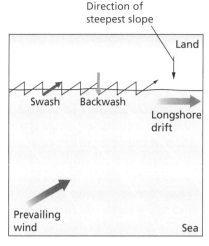

Dispersed energy

Concentrated energy on flanks of headland

Direction of steepest slope

Land

Shoreline

Depth decreasing onshore

Wave movement is slowed down in shallow water, and this causes the wave front to break

Unimpeded fast movement in deep water causes waves to break parallel to each other

Wind direction

Sea

Land

A1   A1

B1   B1

Wave front

A   A   B   B

Sea

Land

Swash   Backwash

Longshore drift

Prevailing wind

Sea

Figure 2.1.1 Wave refraction and longshore drift

1 Outline the processes of weathering and mass movements that occur in coastal environments.
2 Distinguish between swash and backwash.
3 Contrast constructive and destructive waves.

Revision activity

Draw a spider diagram to show the range of processes operating on coastal environments.

Exam tip

Remember that the waves affecting an area of coastline can change seasonally, e.g. destructive waves in winter and constructive waves in summer.

Influences on coastal environments   REVISED ⬤

Coastal environments are influenced by many factors, including physical and human processes. As a result, there is a great variety in coastal landscapes. For example, landscapes vary on account of:

+ **Geology (rock type)** – hard rocks such as limestone give rugged landscapes, whereas soft rocks such as sands and gravels produce low, flat landscapes.
+ **Geological structure** – accordant (or Pacific-type) coastlines occur where the geological strata lie parallel to the coastline, e.g. along the Californian coastline, USA; whereas discordant (or Atlantic-type) coastlines occur where the geological strata are at right angles to the shoreline, e.g. the south-west coastline of Ireland.
+ **Processes** – erosional landscapes, e.g. the east coast of England, contain many rapidly retreating cliffs, whereas areas of rapid deposition, e.g. the Netherlands, contain many sand dunes and coastal flats.

Exam tip

Remember that sea level changes vary around the world due to changes in the level of the land, as well as changes in the level of different seas.

Figure 2.1.2 A raised beach, Portland Island, UK, formed by falling sea levels leaving a beach stranded

Figure 2.1.3 Milford Sound, New Zealand, formed by rising sea level drowning a glaciated valley

+ **Sea-level changes** interact with erosional and depositional processes to produce advancing coasts (those growing either due to deposition and/or a relative fall in sea level) or retreating coasts (those being eroded and/or drowned by a relative rise in sea level). Falling sea levels produce relict cliffs and raised beaches (Figure 2.1.2), whereas rising sea levels are associated with fjords (Figure 2.1.3) and rias (drowned river valleys). An isostatic change is a local change in the level of the land relative to the sea, whereas a eustatic change is a global change in sea level.
+ **Human impacts** are increasingly common – coastal zones are used for many activities including settlement, industry, recreation and tourism, energy developments and transport.
+ **Vegetation/ecosystem type** such as mangrove, coral, sand dune, salt marsh and rocky shore add further variety to the coastline through their impact on microclimate, weathering and their indirect impact on human activities.

Now test yourself

4 Briefly outline how rock type affects coastal landform development.

5 Outline how sea level changes lead to changes in coastal landforms.

6 Briefly explain how vegetation may influence coastal environments.

TESTED ◯

Coastal landforms

REVISED ◯

Features of erosion

On a large scale, bays may be eroded in beds of weaker rock (Figure 2.1.4). The harder rocks form headlands that protrude whereas the weaker rocks are eroded to form bays.

Bay A wide, gently curving indentation of the sea into land.

Headland A piece of high land (promontory) with steep cliffs projecting into the sea.

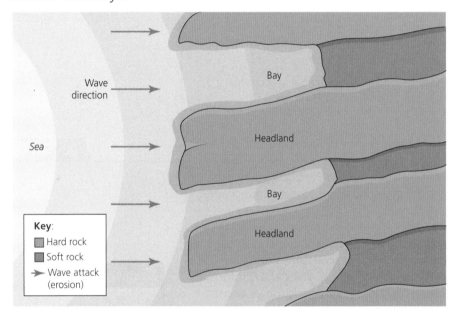

Figure 2.1.4 Headlands and bays

Figure 2.1.5 Headlands and bays in Pelion, Greece.

Answers to Now Test Yourself and Exam Practice questions at **www.hoddereducation.co.uk/myrevisionnotesdownload**

On a headland, erosion will exploit any weakness, creating, at first, a cave. Once the cave reaches both sides of the headland, an arch is formed. A collapse of the top of the arch forms a stack, and when the stack is eroded a stump is created (Figure 2.1.6).

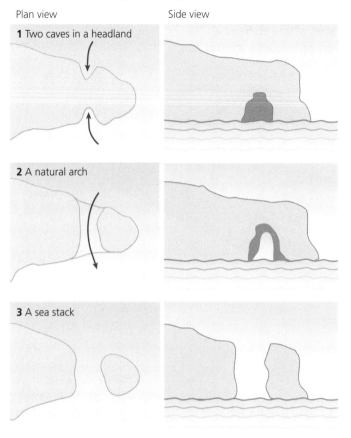

1 Wave refraction concentrates erosion on the sides of headlands. Weaknesses such as joints or cracks in the rock are exploited, forming caves.
2 Caves enlarge and are eroded further back into the headland until eventually the caves from each side meet and an arch is formed.
3 Continued erosion, weathering and mass movements enlarge the arch and cause the roof of the arch to collapse, forming a high standing stack.

Figure 2.1.6 Formation of caves, arches, stacks and stumps

A wave-cut platform is a relatively flat, long slope, found at the base of some cliffs formed by wave action.

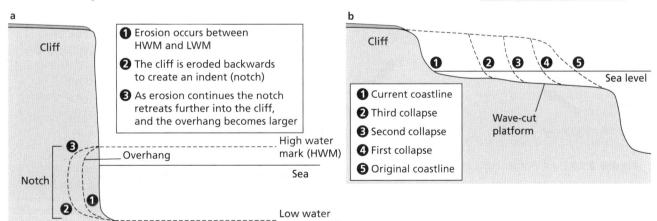

Figure 2.1.7 Formation of a wave-cut platform

Exam tip

Remember that most erosion happens very slowly – it is only during extreme storm events that most erosional features are formed.

Revision activity

Draw a spider diagram to list as many coastal erosional landforms as you can.

Wave cut platform A gently sloping area that extends from the base of a cliff. They are formed by marine erosion although weathering may also help from the platform.

Features of deposition

Deposition occurs with constructive waves, a good supply of material and a sheltered location.

Beaches

A beach is a deposit of sand or shingle formed in an area where there is a large supply of material, constructive waves and/or strong onshore winds that carry sediments inwards during low tides.

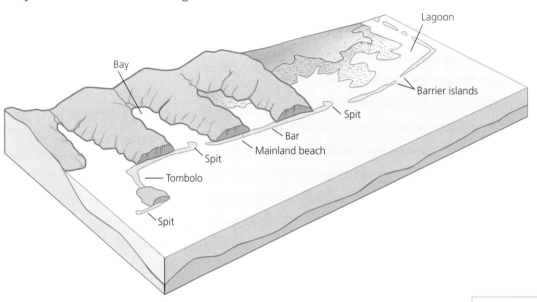

Figure 2.1.8 Features of coastal deposition

A spit is a beach of sand or shingle linked at one end to land (Figure 2.1.8). It is found where wave energy is reduced, e.g. along a coast where headlands and **bays** are common and near river mouths. Spits often become curved as waves undergo refraction (Figure 2.1.9).

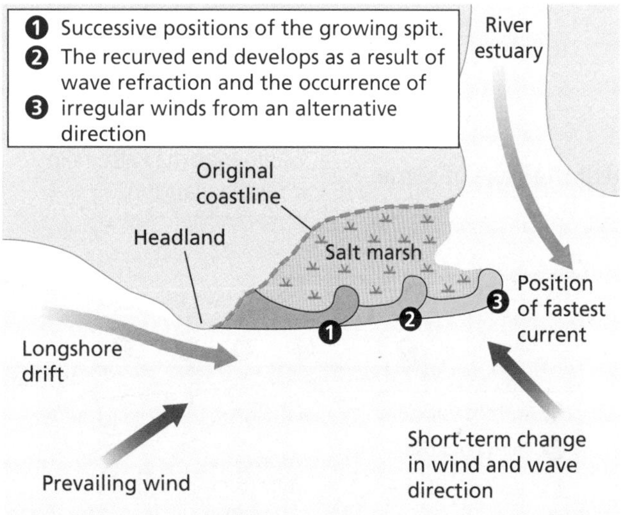

① Successive positions of the growing spit.
② The recurved end develops as a result of wave refraction and the occurrence of
③ irregular winds from an alternative direction

River estuary

Original coastline

Headland

Salt marsh

② ③ Position of fastest current
①

Longshore drift

Prevailing wind

Short-term change in wind and wave direction

Figure 2.1.9 Development of a spit

Related features include bars. These are ridges that block off a bay or river mouth.

Beach An accumulation of sand/shingle that may occur in sheltered areas or in exposed areas where there is a plentiful supply of sediment.

Spit A ridge of sand or shingle that is connected to the mainland at one end but the other end is in the ocean.

Tombolo A ridge of sand or shingle that is that connects an island to the mainland.

Bar An accumulation of sand or shingle that extends across a bay between two headlands (a bay bar) or deposited in water offshore parallel to the coast (offshore bar).

Prevailing wind The direction of the most frequent wind in an area.

7   In your own words, describe how a wave-cut platform may be formed.
8   Identify feature A shown in Figure 2.1.10, and explain how it may have been formed.

Figure 2.1.10 A coastal feature – some of the Twelve Apostles (now reduced to eleven), Australia

9   Outline the main characteristics of spits.

Revision activity

Make a list of the main conditions needed for depositional landforms to develop.

Exam tip

Increasingly human activities are affecting coastal processes and landforms, especially depositional ones.

2.2 Distinctive ecosystems develop along particular areas of coastline

Coastal ecosystems

REVISED

There are several types of coastal ecosystems including coral reefs and mangroves, sand dunes and salt marshes.

Coral reefs

Coral requires sea surface temperatures (SSTs) of 17–33°C for growth, salinity levels of 30–38 parts per thousand and clear water. **Coral** reefs have a rich biodiversity – about 25% of the world's sea fish breed, grow, spawn and evade predators in coral reefs.

Coral reef A reef composed of limestone due to the accumulation of coral. When coral die they leave behind a hard skeleton of calcium carbonate which appears as rock.

Mangroves Salt-tolerant forests that grow in tidal estuaries and muddy coastlines of tropical areas.

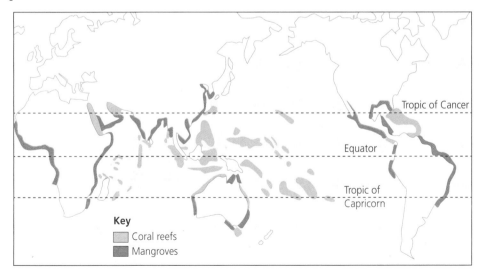

Figure 2.2.1 The global distribution of coral reefs and mangroves

Mangroves

Mangroves are salt-tolerant forests that grow in the tidal estuaries and muddy coastlines of tropical areas. They require SSTs of over 24°C in the warmest months and annual rainfall of over 1250mm. The muddy waters, rich in nutrients from decaying leaves and wood, are home to a large biodiversity.

Sand dunes

Sand dunes are common in storm wave environments. They are favoured in areas where there is a prevailing onshore wind, a large supply of sand and a large tidal range.

Conditions on sand dunes are saline, windy and dry. Vegetation which dominates sand dunes is salt-tolerant (halophytic) and xerophytic (drought-resistant) grasses.

Halophytes Plants that are adapted to salty conditions.

Xerophytes Plants that are adapted to dry conditions.

Salt marshes

Salt marshes are very productive and fertile ecosystems. They are found in sheltered locations, e.g. behind a spit and/or in tidal river estuaries. They have a high oxygen content, high nutrient and light availability. Salt marsh vegetation is halophytic. It also has deep roots to anchor the plant in the mud and the ability to extract nitrogen directly from the air.

A thin layer of mud forms over sand which is covered at each tide. The only plants are algae growing on the mud.

More mud deposited and the first plants appear. The plants trap more mud and silt. The marsh is covered at each high tide and channels are cut as the water recedes.

Salicornia herbacea — Spartina townsendii

Glyceria maritima — Halimione portulacoides

Further plants appear higher up the marsh. This accelerates mud accretion. Channels deepen as the marsh surface rises.

Armeria maritima — Limonium vulare

More plants move into the higher zones and the mud deepens. High tides still flood the marsh but low tides are confined to the creeks, which are further eroded as the water runs off.

Juncus maritimus — Festuca rubra

The marsh is now growing slowly and the mud is very deep. Further plants colonise the higher zones. Erosion undercuts the creek banks and some collapse leaving bare salt pans above the collapse. Apart from creeks and pans, the marsh is covered with vegetation and only the highest tides fully cover it with water.

Figure 2.2.2 Salt marsh formation

Exam tip

Use specialist terms such as halophytic and xerophytic to describe vegetation.

Revision activity

Draw a diagram of a salt marsh, showing different plant zones within it.

Now test yourself TESTED

1 Compare and contrast the distribution of coral reefs with that of mangroves.
2 Contrast the conditions needed for the formation of salt marshes with those needed for the development of sand dune ecosystems.
3 Suggest possible reasons why there are no coral reefs off the west coast of South America.

Abiotic and biotic characteristics of a named coastal ecosystem

All ecosystems include biotic (living) and abiotic (non-living) components. Sand dunes vary from place to place, but many show similar changes with distance from the sea in terms of their biotic and abiotic components.

> **Biotic** Living elements in an ecosystem.
>
> **Abiotic** Non-living elements in an ecosystem.

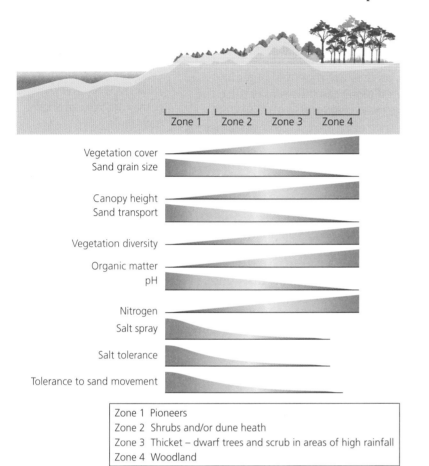

Zone 1 Pioneers
Zone 2 Shrubs and/or dune heath
Zone 3 Thicket – dwarf trees and scrub in areas of high rainfall
Zone 4 Woodland

Figure 2.2.3 Abiotic and vegetation gradients across a typical coastal sand dune

> **Revision activity**
>
> Look back at Section 2.2 and identify the environmental conditions on sand dunes.

> **Exam tip**
>
> Not all sand dune ecosystems show the same changes shown in the diagrams above. It depends, in part, on the climate, geology, width of dunes, human activity, salinity and pH of the area.

> **Now test yourself**
>
> TESTED ○
>
> 4 Describe how abiotic characteristics change with distance from the shoreline.
> 5 State how biotic characteristics change with distance from the shoreline.
> 6 Suggest how and why wind speed will vary from the seashore to the back of the dunes.

Threats to coastal ecosystems

There are many threats to coastal ecosystems including industrialisation, agriculture, tourism and deforestation.

Coral reefs

Physical damage to reefs may be caused by quarrying, dredging, propellers and anchors, destructive fishing practices (using explosives), removal for souvenirs and 'development' in general. In addition, coral is very vulnerable to pollution. Run-off from fertilisers or sewage can be particularly damaging as coral reefs are adapted to low nutrient levels. Over-fishing may reduce the number of grazing fish that keep coral clear of algae. Fishing using explosives may damage reefs.

Coral reefs are also threatened by global climate change. Oceanic warming causes the coral to expel the microscopic algae that produce the food that coral needs, leading to coral bleaching.

Ocean acidification refers to the increasing amount of carbon dioxide in seawater, causing seawater to become more acidic. This has a negative impact on any shell-forming organism, as well as on coral reefs.

Mangroves

There are many threats to mangroves including clearance for coastal developments, e.g. hotels; deforestation for wood and charcoal; browsing by grazers; human trampling and souvenir collection; over-fishing in adjacent areas; waste from nearby settlements and human activities; pollution, e.g. oil spills; and changes to local landscapes, e.g. coastal jetties. Mangroves may also be removed for the control of malaria, as they are potential breeding grounds for mosquitoes. Power plants located along coastlines may raise the temperature of seawater and cause damage to ecosystems.

Now test yourself TESTED ◯

7 Briefly explain the impacts of global climate change on coral reefs.

8 Suggest why mangroves are vulnerable to human activities.

9 Suggest why coral reefs attract many tourists.

> **Revision activity**
>
> Draw a spider diagram to show the ways in which human activity can affect coral reefs.

> **Exam tip**
>
> Try to use named examples in examination responses.

Sand dunes

Human activities can affect sand dunes in many ways including disruption of sediment flow, coastal development, recreational use, grazing, introduction of exotic species and sand mining. The construction of groynes and sea walls may disrupt sediment movement.

Sand dunes are used intensively for recreation. Vegetation may be trampled by people. Some dunes allow four-wheeled drive vehicles and horse-riding. The weight of these activities compacts the soil. Compacted soil has less oxygen so vegetation growth is reduced. Some dunes have been flattened for sports activities and for car-parking.

Sand is also mined. It is one of the most important resources in the construction industry, used in cement and concrete. Coastal and river sand is better for construction than desert sand, as it is coarser and binds together better.

Salt marshes

Salt marshes are extremely vulnerable to coastal development. Waterfront property is very valuable and much sought after. However, development can lead to an increase in noise and light pollution and may affect wildlife behaviour and reproductive success.

Salt marshes are vulnerable to eutrophication caused by excess nitrogen and phosphorus. Industries and vehicles may release heavy metals which may damage food chains in salt marshes. Pesticides and insecticides may also end up in salt marshes. Salt marshes may store pollutants and toxins for a long time.

Roads have often divided salt marshes into two sections. In cases where tidal flooding has been eliminated, invasive species, such as the common reed, have become established. Elsewhere salt marshes have been drained in attempts to control mosquitoes and this can lead to a decline in biodiversity.

> **Eutrophication** Pollution of water due to high levels of nitrates (in fertilisers) and/or phosphates (in detergents).

> **Revision activity**
>
> Draw two spider diagrams – one showing human activities on sand dunes, the other showing human activity on salt marshes.

> **Exam tip**
>
> There are many conflicts over the use of sand dunes and salt marshes. There are pressures for recreation, conservation, development and housing.

Now test yourself TESTED ◯

10 Suggest why demand for sand mining is increasing.

11 Outline the impacts of coastal activities on salt marshes.

12 Outline reasons for the conservation of sand dunes and/or salt marshes.

2.3 Coastal environments are of great importance to people and need to be sustainably managed

Conflicts between different users

REVISED ○

There are many different users in coastal areas and this can lead to conflict. For example, tourists, fishermen, industrialists and farmers may have different needs from the same coastal region.

Exam tip

Try to have a supporting example/case study to support your comments about coastal conflicts.

Table 2.3.1 Relationships between human activities and coastal zone problems

Human activity	Possible consequences	Potential coastal zone problems
Urbanisation and transport	Land-use changes (e.g. for ports, airports); congestion; water abstraction; waste disposal	Loss of habitats and species diversity; water pollution; introduction of alien species
Agriculture	Land reclamation; fertiliser and pesticide use; livestock densities; water abstraction	Loss of habitats and species diversity; water pollution; eutrophication
Tourism and recreation	Development and land-use changes; congestion; ports and marinas; water abstraction; waste disposal	Loss of habitats and species diversity; disturbance; lowering of groundwater table; water pollution
Fisheries and aquaculture	Port construction; fish-processing facilities; fishing gear; fish farm pollution	Overfishing; litter and oil on beaches; introduction of alien species; habitat damage
Industry (including energy production)	Land-use changes; power stations; extraction of natural resources; process effluents; cooling water	Loss of habitats and species diversity; water pollution; eutrophication; thermal pollution

An example of the successful management of conflicting land uses is the Soufrière Marine Management Area in St Lucia. This uses a system of land-use zoning to allow different activities to occur in designated areas (Figure 2.3.2).

Revision activity

Create a spider diagram to identify potential conflicts over the use of the Barcelona coastline, as shown in Fig. 2.3.1.

Figure 2.3.1 Part of the Barcelona coastline.

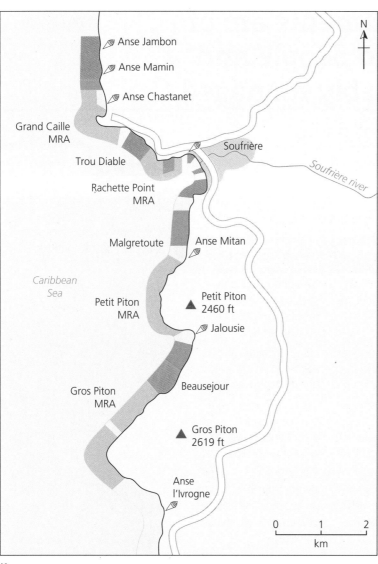

Key

■ **Marine reserves**
Allows fish stocks to regenerate and protects marine flora and fauna.
Access to the area is by permit and can be enjoyed by divers and snorkellers.

□ **Fishing priority areas**
Commercial fishing has precedence over all other activities in these areas.

■ **Yachting areas**
Yachting is not allowed in the SMMA. Moorings are provided in these areas only.

■ **Multiple-use areas**
Fishing, diving, snorkelling and other legitimate uses are allowed.

⬚ **Recreational areas**
Areas for public recreation – sunbathing, swimming.

Figure 2.3.2 Land-use zoning in the Soufrière Marine Management Area (SMMA), St Lucia

Now test yourself

TESTED ◯

1 Suggest a consequence and a potential problem if 'Conservation' was a category in Table 2.3.1.

2 Identify the varied users (**stakeholders**) in the Soufrière Marine Management Area.

3 Explain how fishing and aquaculture can lead to problems in coastal areas.

Revision activity

Make a list of the different users in Figure 2.3.1 and suggest which other groups each one is in conflict with.

Exam tip

When a scale is given on a map, make sure you use it. How large is the Soufrière Marine Management Area in Figure 2.3.1?

Stakeholders Any users of a coastal zone with a particular interest, e.g. residents, tourists, fishermen.

Causes of coastal flooding

Coastal floods are caused by many factors including storm surges, tsunamis and, increasingly, global climate change. Often the worst floods are a combination of these.

A storm surge is a rapid rise in sea level caused by high wind speeds. Tsunamis are large sea waves caused by submarine earthquakes. Tsunamis that occur close to the coastline have greater impact than those generated further out at sea. Climate change is leading to increasing frequency and magnitude of coastal flooding due to increased atmospheric energy.

Prediction

Coastal flood forecasts are concerned with high magnitude waves, high tides and/or storm surges, occurring either separately or together.

Most storms are tracked by satellites, especially by high-income countries (HICs), and predictions are made about their likely path. It is difficult to give much notice about tsunamis given their speed although there is a Pacific Tsunami Warning System.

Improved forecasting is allowing many communities to take evasive action (evacuate or take shelter) ahead of coastal flooding. Early warning systems in Bangladesh monitor tropical storms and monsoon rains and provide regular updates on the paths of storms and their likely impacts.

Building design

The main aim of coastal flood engineering is to prevent coastal erosion and flooding. Two main approaches have been used. One is to elevate buildings, so that floodwaters may pass under the building. The second is to flood-proof buildings through raised foundations, reinforced barriers, dry flood proofing (sealing a property so that floodwaters cannot enter) and wet flood proofing (allowing partial flooding of buildings).

Exam tip

Remember that most coastal protection will protect against an event of a certain size, e.g. the size of flood you might expect once a century. In addition, flooding is becoming more frequent and larger in size due to global climate change and some cities gradually sinking (subsiding).

Now test yourself TESTED

4 Suggest why it is difficult to warn against tsunamis.
5 Explain how coastal buildings may be made 'flood proof'.
6 Identify the main causes of coastal flooding.

Coastal management strategies

Shoreline management plans

A shoreline management plan (SMP) is an attempt to protect an area of coastline without leading to problems elsewhere. The coast is divided into sediment cells, i.e. natural units, and for each cell local governments can decide how best to manage the coast. Some areas may be allowed to be eroded whereas others may be given protection.

Integrated coastal zone management

Integrated coastal zone management (ICZM) is an attempt to manage all aspects of a coastal system, e.g. marine areas, land, people and economic activities. It tries to balance protecting the coastline with its use by people and the economy. ICZM considers the interdependence of marine and terrestrial systems, different users (stakeholders) and different scales, e.g. local and national importance.

Hard engineering structures

Hard engineering structures include groynes, sea walls, revetments, rock armour and cliff drains (see Table 2.3.2). They try to alter natural processes so as to reduce the potential for erosion of the coastline. Sometimes they may have unexpected results, e.g. the building of groynes may reduce erosion in nearby locations but increase coastal erosion down drift. Sea walls may also lead to the scouring of the seabed.

Soft engineering

Soft engineering refers to working with nature. Examples include the maintenance of a mangrove forest to reduce the impact of tropical storms. Beach nourishment increases the size of a beach by using sediment dredged from elsewhere. Some soft engineering may allow the coastline to retreat naturally.

Managed retreat allows nature to take its course – erosion in some areas, deposition in others. Benefits include less money being spent and the creation of natural environments. However, some homes or farms may be lost to the power of the sea.

> **Hard engineering**
> Attempts to manage coastal areas by altering natural processes through the use of man-made structures.
>
> **Soft engineering**
> Attempts to manage coastal erosion/flooding by working with nature, e.g. afforestation.

Table 2.3.2 Hard and soft engineering structures

Type of management	Aims/methods	Strengths	Weaknesses
Hard engineering	**To control natural processes**		
Cliff base management	**To stop cliff or beach erosion**		
Sea walls	Large-scale concrete curved walls designed to reflect wave energy	Easily made; good in areas of high density	Expensive; life span about 30–40 years; foundations may be undermined
Revetments	Porous design to absorb wave energy	Easily made; cheaper than sea walls	Life span limited
Gabions	Rocks held in wire cages to absorb wave energy	Cheaper than sea walls and revetments	Small-scale
Groynes	To prevent longshore drift	Relatively low costs; easily repaired	May cause erosion on downdrift side; interrupts sediment flow
Rock armour (rip rap)	Large rocks at base of cliff to absorb wave energy	Cheap	Unattractive; small-scale; may be removed in heavy storms
Cliff face strategies	**To reduce the impacts of subaerial processes**		
Cliff drainage	Removal of water from rocks in the cliff	Cost-effective	Drains may become new lines of weakness; dry cliffs may produce rock falls
Cliff regrading	Lowering of slope angle to make cliff safer	Useful on clay (most other measures are not)	Uses large amounts of land – impractical in heavily populated areas
Soft engineering	**Working with nature**		
Beach replenishment	Sand pumped from seabed to replace eroded sand	Looks natural	Expensive; short-term solution
Ecosystem rehabilitation and revegetation	Restoring coastal ecosystems	Natural	Can take a long time to achieve; may be subject to other pressures for development
Managed retreat	Coastline allowed to retreat in certain places	Cost-effective; maintains a natural coastline	Unpopular; political implications
'Do nothing'	Accept that nature will win	Cost-effective!	Unpopular; political implications

Figure 2.3.3 Coastal management 1

Figure 2.3.4 Coastal management 2

Exam practice

1 Identify two types of mass movement that are important in coastal reas. (1 mark)

2 Distinguish between isostatic and eustatic changes in sea level. (2 marks)

3 Describe the distribution of coral reefs as shown in Figure 2.2.1. (2 marks)

4 Describe how vegetation varies across a sand dune ecosystem. (4 marks)

5 Explain how spits are formed. (4 marks)

6 Explain the advantages and disadvantages of land-use zoning as a form of coastal management. (4 marks)

7 Outline the advantages and disadvantages of hard engineering methods of coastal management. (8 marks)

Total: 25 marks

Now test yourself

7 Distinguish between hard and soft engineering.

8 Distinguish between integrated coastal zone management and shoreline management plans.

9 Briefly explain how the management schemes in Figures 2.3.3 and 2.3.4 function.

TESTED ◯

Summary

✚ There are many physical processes operating on coastlines including marine processes of erosion and deposition, weathering and mass movements.

✚ Many factors influence the nature of coastal environments including geology, vegetation, sea level changes and human activities.

✚ Erosional and depositional processes are important for the formation of landforms such as headlands and bays, cliffs, stacks, beaches and spits.

✚ Many coastal areas have distinctive ecosystems such as coral reefs, mangroves, sand dunes and salt marshes.

✚ Coastal ecosystems have biotic and abiotic components.

✚ Coastal ecosystems are vulnerable to human activities.

✚ There are many conflicts between different users of coastal areas.

✚ Coastal flooding can be caused by storm surges, tsunamis and climate change, although flooding may be predicted and attempts made to prevent flooding of buildings.

✚ Coastal management strategies can be divided into hard and soft engineering.

Exam tip

Remember that many stretches of coastline have a range of management types – usually these will be a mix of hard and soft engineering, often side by side.

3 Hazardous environments

3.1 Natural hazards

Characteristics, distribution and measurement

A natural hazard is defined as an extreme event or condition in the natural environment causing harm to people, properties or livelihoods. Natural hazards only lead to natural disasters because people live in hazardous areas. There are several types of natural hazards.

Table 3.1.1 Different types of natural hazards

Tectonic and geological	Climatic and meteorological
Earthquakes	Tropical cyclones
Volcanoes	Drought
Tsunamis	Floods
Landslides	Tornadoes

It is possible to characterise hazards and disasters in several ways:
+ Magnitude: the size of the event, e.g. the size of an earthquake on the Moment Magnitude Scale.
+ Frequency: how often an event of a certain size occurs. The frequency is sometimes called the recurrence interval – the larger the event, the less frequently it occurs. However, it is the very large events that do most of the damage (to the physical environment, to people, properties and livelihoods).
+ **Regularity**: some hazards, such as tropical cyclones, are regular, whereas others, such as earthquakes and volcanoes, are much more random.
+ **Areal extent**: the size of the area covered by the hazard.
+ **Spatial concentration/dispersion** is the distribution of hazards over space, whether they are concentrated in certain areas, such as tectonic plate boundaries, coastal locations, valleys and so on.
+ **Speed of onset** varies from very rapid events, such as an earthquake, to slower timescale events such as tropical cyclones over a period of many days.
+ **Duration**: the length of time that a natural hazard exists.

Magnitude The size of the event.

Moment Magnitude Scale A scale that measures the amount of energy released during an earthquake.

Frequency How often an event of a certain size occurs.

Cyclone An area of low pressure, characterised by strong winds and heavy rainfall.

Eye The calm, central part of a tropical cyclone.

Tropical cyclones (hurricanes)

Tropical cyclones are intense, low-pressure systems that bring heavy rainfall, strong winds and high waves, and cause other hazards such as flooding and mudslides. They are large-scale features with a diameter of up to 800 km and a calm central area, the eye. Most hurricanes take place in tropical and sub-tropical regions.

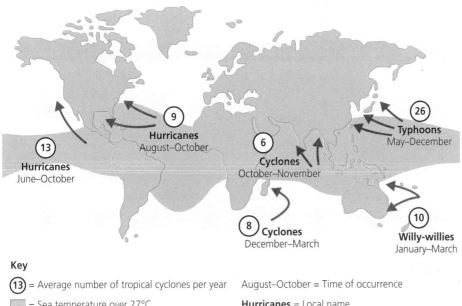

Key

(13) = Average number of tropical cyclones per year   August–October = Time of occurrence

▨ = Sea temperature over 27°C   **Hurricanes** = Local name

Figure 3.1.1 Distribution of tropical cyclones

Table 3.1.2 The Saffir–Simpson Scale

Hurricane category	Pressure (mb)	Wind speed (kmph)	Approx. storm surge (m)
1	≥980	119–153	1.2–1.5
2	965–979	154–177	1.6–2.6
3	945–964	178–209	2.7–3.6
4	920–944	210–251	3.7–5.5
5	<920	≥252	Over 5.5

Earthquakes

An earthquake is a sudden, violent shaking of the Earth's surface. Earthquakes occur after a build-up of pressure causes rocks and other materials to give way. Most of this pressure occurs at plate boundaries when one plate is moving against another. Earthquakes are associated with all types of plate boundaries. Broad belts of earthquakes are associated with subduction zones, whereas narrower belts of earthquakes are associated with constructive plate margins. Collision boundaries are also associated with broad belts of earthquakes, whereas conservative plate boundaries give a relatively narrow belt of earthquakes.

The focus refers to the place beneath the ground where the earthquake takes place. **Deep-focus earthquakes** are associated with subduction zones. **Shallow-focus earthquakes** are generally located along constructive boundaries and along conservative boundaries. The epicentre is the point on the ground surface immediately above the focus.

Measurement

The Richter scale was developed in 1935 to measure the magnitude of earthquakes. The scale is logarithmic, so an earthquake of 5.0 on the Richter scale is ten times more powerful than one of 4.0 and one hundred times more powerful than one of 3.0. Scientists are increasingly using the Moment Magnitude Scale (M) which measures the amount of energy released and produces figures that are similar to the Richter scale. For every increase of 1.0 on the M scale, the amount of energy released increases by over 30 and an increase of ten times the amplitude recorded by a seismograph.

> **Focus** The exact position where an earthquake takes place.
>
> **Epicentre** The position on the ground surface directly above the focus.
>
> **Richter scale** A scale that measures the magnitude of an earthquake.

Table 3.1.3 Annual frequency of occurrence of earthquakes of different magnitude based on observations since 1900

Descriptor	Magnitude (Richter scale)	Average number/years	Hazard potential
Great	>8	1	Total destruction, high loss of life
Major	7.0–7.9	18	Serious building damage, major loss of life
Strong	6.0–6.9	120	Large losses, especially in urban areas
Moderate	5.0–5.9	800	Significant losses in populated areas
Light	4.0–4.9	6200	Usually felt, some structural damage
Minor	3.0–3.9	49,000	Typically felt but usually little damage
Very minor	<3.0	9000/day	Not felt, but recorded

Volcanoes

A volcano is an opening through the Earth's crust through which hot molten magma and ash are erupted onto the land as lava, ash and cinders. Most volcanoes are found at plate boundaries (constructive and destructive/subduction zones) although some occur over hotspots. About three-quarters of the Earth's 550 historically active volcanoes lie along the Pacific Ring of Fire.

The Hawaii volcanoes are found in the middle of the ocean and occur at a hotspot. A hotspot is a plume of hot material rising from deep within the mantle which is responsible for the volcanoes.

> **Magma** Molten material from the Earth's interior.
>
> **Lava** Molten material from the Earth's interior ejected at the Earth's surface through a volcano or a crack.

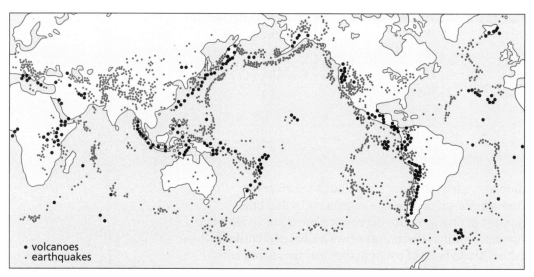

• volcanoes
· earthquakes

Figure 3.1.2 Distribution of volcanoes and earthquakes

The strength of a volcano is measured by the volcanic explosive index (VEI). This is based on the amount of material ejected in the explosion, the height of the cloud it causes, and the amount of damage. Any explosion above level 5 is very large and violent. A VEI 8 is also known as a supervolcano.

> **Cone volcano** Steep-sided volcano by slow moving, viscous (sticky) lava.
>
> **Shield volcano** Low angle volcano, formed by very hot, runny lava.

Table 3.1.4 Magnitude and frequency of volcanic eruptions

VEI	Volume of materials erupted	Frequency of eruption	Example	Occurrence in last 10,000 years
0	<10,000 m³	Daily	Kilauea	Many
1	<100,000 m³	Daily	Hekla, Iceland	Many
2	1,000,000 m³	Weekly	Unzen, Japan (1990)	c. 3500
3	10,000,000 m³	Yearly	Nevado del Ruiz (1985)	c. 900
4	0.1 km³	>10 years	Soufrière Hills (1995)	c. 275

→

VEI	Volume of materials erupted	Frequency of eruption	Example	Occurrence in last 10,000 years
5	1 km³	>50 years	Mt St Helens (1980)	c. 85
6	10 km³	>100 years	Krakatoa (1883)	c. 40
7	>100 km³	>1000 years	Tambora (1815)	4
8	>1000 km³	>10,000 years	Toba (73,000 BC)	None

Now test yourself · TESTED

1 Identify the region with the greatest number of tropical cyclones per year.
2 Briefly explain the location of the world's volcanoes.
3 How much more powerful is an earthquake of magnitude 7.0 compared with one of magnitude 4.0?

Causes of tropical cyclone hazards

REVISED

Tropical cyclones develop from low pressure systems. They originate over oceans that have sea surface temperatures of over 27°C in order for sufficient evaporation to occur. They develop away from the Equator as there is insufficient rotation (Coriolis Force) there. For a tropical cyclone to form, lower and upper winds need to be coming from the same direction. This means that vertical wind shear (the change in wind speed and direction with height) is reduced.

Tropical storms bring intense rainfall and very high winds, which may in turn cause storm surges and coastal flooding, and other hazards such as flooding and mudslides. The impact of different strength of tropical cyclones is shown in Table 3.1.5.

Table 3.1.5 The impact of tropical cyclones

Category	Description
Category 1: Winds 119–153 km/h; storm surge generally 1.2–1.5 m above normal	No real damage to building structures. Damage primarily to unanchored mobile homes. Also, some coastal road flooding and minor pier damage.
Category 3: Winds 178–209 km/h; storm surge generally 2.7–3.6 m above normal	Some structural damage to small residences and utility buildings. Mobile homes are destroyed. Flooding near the coast destroys smaller structures, with larger structures damaged by floating debris. Evacuation of low-lying residences close to the shoreline may be necessary.
Category 5: Winds greater than 252 km/h; storm surge generally greater than 5.5 m above normal	Complete roof failure on many residences and industrial buildings. Some small buildings blown over or blown away. Low-lying escape routes are cut by rising water 3–5 hours before arrival of the centre of the tropical storm. Massive evacuation of residential areas on low ground within 8–16 km of the shoreline may be required.

Now test yourself · TESTED

4 Identify the main conditions needed for the formation of tropical cyclones.
5 State the estimated wind speed and storm surge associated with a category 3 tropical cyclone.
6 Outline the likely damage to buildings caused by a category 5 tropical cyclone.

Volcanic and earthquake hazards

Volcanic hazards occur by constructive and destructive plate margins and near hot spots. They do not occur near conservative plate margins or collision zones. The underlying cause of volcanic hazards is the ejection of magma from within the Earth onto the Earth's surface. However, the nature of volcanic eruptions varies. At constructive margins, associated with rising magma within the Earth, the lava is hot and runny and produces less explosive eruptions. In contrast, at destructive plate margins, the lava is acidic and sticky and produces explosive eruptions which can include pyroclastic flows and lava flows. Destructive eruptions are more likely to lead to acidification and climate change as they put more material into the atmosphere. If volcanic ash mixes with water it can create lahars or mudflows. Lava flows at hot spots, e.g. Hawaii, are relatively slow moving but they can burn buildings and vegetation. However, volcanic hazards are only hazards when people are near to volcanoes.

There are a number of primary hazards and secondary hazards related to volcanic eruptions. The impacts will depend on the magnitude of the event and the population at risk.

Hot spots Isolated, rising plumes of magma that lead to volcanic activity away from plate margins.

Pyroclastic flows Explosive clouds of superheated material (up to 700 °C) that can travel at up to 500 km/hour.

Acidification The increased acidity of water due to sulphur emissions from volcanoes.

Lahars Mudflows caused by the mixing of volcanic ash and water.

Primary hazards The direct hazards associated with natural events.

Secondary hazards The indirect hazards associated with natural events.

Liquefaction The way in which soil, loose materials and some rocks act like a liquid due to shaking during an earthquake.

Table 3.1.6 Primary and secondary hazards associated with volcanoes

Primary hazard	Secondary hazards
Lava flows	Lahars (mudflows)
Ash fallout	Landslides
Pyroclastic flows	Acidification
Gas emissions	Climate change (global cooling)
	Fire

Earthquake hazards

Earthquakes commonly occur at plate boundaries. There is a build-up of pressure as plates move in different directions. The release of pressure results in a sudden movement which is an earthquake. Some earthquakes may be caused by human activity such as mining, building of large dams and the underground testing of bombs.

Table 3.1.7 Primary and secondary hazards associated with earthquakes

Primary hazards	Secondary hazard
Ground shaking	Ground failure and soil liquefaction
	Building collapse
	Gas leaks and fires
	Landslides and rock falls
	Debris flow and mudflow
	Tsunamis

Revision activity

Make a list of the different types of plate boundary, and state whether they are associated with volcanoes, earthquakes or both.

Now test yourself

TESTED

7 Distinguish between a lahar and a pyroclastic flow.
8 Distinguish between primary and secondary hazards.
9 Suggest why secondary hazards related to earthquakes kill more people than primary hazards.

Exam tip

Remember that not all types of secondary hazards occur in every volcanic eruption or earthquake event.

Answers to Now Test Yourself and Exam Practice questions at **www.hoddereducation.co.uk/myrevisionnotesdownload**

3.2 Hazards have an impact on people and the environment

Reasons why people live in hazardous environments

REVISED

Why do people often live in hazardous environments?

Natural hazards occur only when people, livelihoods and/or property are at risk. Although some causes of hazards may be tectonic or climatic, it is because people live in such areas that makes them hazardous. If no one was affected, it would not be a hazard.

Why do people live in such places?

+ Some people consider that the potential advantages of living in an area outweigh the potential risks.
+ Another view is that poor people have little choice in where they live. Hence, poor people live in unsafe areas – such as steep slopes or floodplains – because they are prevented from living in better areas.

For example, deltas provide water, silt, fertile soils and the potential for trade and communications. They may also be subject to tropical cyclones, as shown by the 2020 floods caused by Cyclone Amphan in the Ganges Delta.

> **Risk** The probability of a natural hazard causing harmful consequences, e.g. loss of life, injury, damage to properties, the economy and/or the environment.
>
> **Vulnerability** The geographic conditions that increase the susceptibility of a community to a natural hazard.

Figure 3.2.1 View of part of Mt Etna – a source of both risk and economic potential

> **Revision activity**
>
> Make two spider diagrams – one showing the benefits and the other showing the disadvantages of living in tropical coastal areas.

> **Now test yourself**　　　　　　　　　TESTED
>
> 1 State two reasons why people live in hazardous areas.
> 2 Referring to Figure 3.2.1, outline why Mt Etna can be considered a hazard and a resource.
> 3 Suggest reasons why people live in areas that are subject to earthquakes.

> **Exam tip**
>
> Remember that hazardous environments can also create employment opportunities – it is up to people to weigh up the risks.

Vulnerability to natural hazards

REVISED

The concept of vulnerability includes not only the physical effects of a natural hazard but also the status of people and property in the affected area. Several factors can increase people's vulnerability to natural hazards.

Table 3.2.1 Factors affecting vulnerability

Economic factors

+ *Levels of wealth and development*: These influence building quality. People in high-income countries (HICs) generally have better quality housing than poorer communities in low-income countries (LICs), especially in slum areas.
+ *Building styles and building codes*: These affect the safely of buildings. Some countries have a more rigorous enforcement of building regulations, e.g. Japan.
+ *Access to technology*: People with access to ICT may have more warnings. Japan sends out text messages to warn people about tsunamis and earthquakes.
+ *Insurance cover*: The poor cannot afford insurance cover. To have insurance cover, buildings need to be made hazard-resistant.

Social factors

+ *Education*: People with a better education generally have a higher income and can afford better quality housing.
+ *Gender*: Many women are carers for their children and/or their parents and they may feel responsible for them following an event.
+ *Population density*: Many rapidly growing cities and large urban areas are especially vulnerable to natural hazards.
+ *Age*: Elderly people and some with disabilities may be far less mobile than younger populations.

Physical factors

+ *Physical geography*: Some areas experience more natural hazards, e.g. along fault lines, near volcanoes and the risk of hurricanes in tropical regions.
+ *Natural environment*: Some areas, e.g. steep slopes and flat, coastal lowlands, are vulnerable to natural hazards.

Table 3.2.2 Number of natural disasters and consequent number of deaths per income group, 1994–2013

Income	Disasters (%)	Number of disasters	Deaths (%)	Number of deaths
High income	26%	1700	13%	182,000
Upper middle-income	30%	1992	19%	252,000
Lower middle-income	27%	1751	35%	474,000
Low income	17%	1119	33%	441,000

Source: Based on data in CRED, 2015.

Revision activity

Using the data in Table 3.2.2, make two pie charts, one showing the percentage of disasters in the four types of country, and the other showing the percentage of deaths.

Exam tip

Remember that some people are more vulnerable to natural hazards than others – poverty is a major factor which increases risk.

Now test yourself   TESTED ⬤

4   Briefly explain why some people are more vulnerable to hazards than others.
5   Outline the main impact of natural hazards on income groups.
6   Explain why natural hazards vary from place to place.

Short-term and long-term impacts   REVISED ⬤

Nepal earthquake, 2015

Short-term impacts

In 2015 there was a 7.8 magnitude earthquake in Nepal.

+ There were over 300 aftershocks, some of them reaching magnitudes of over 7.0.
+ The main earthquake was a shallow-focus earthquake just 80 km from Kathmandu.
+ Rapid population growth in Kathmandu had increased the vulnerability of the area to earthquakes.
+ Nearly 9000 people were killed and 20,000 injured.
+ Overall, 8 million people were affected.
+ Over 600,000 homes were destroyed and over 250,000 homes were damaged.

Answers to Now Test Yourself and Exam Practice questions at **www.hoddereducation.co.uk/myrevisionnotesdownloac**

- Water and electricity were not available in many places following the earthquake.
- The government immediately began to search for people in collapsed buildings.
- Temporary shelters were provided for those made homeless.
- Temporary schools made of bamboo and tarpaulin opened after a month.

Long-term impacts

Longer-term impacts largely related to rebuilding housing.
- One year after the earthquake, towns and villages outside of Kathmandu remained severely damaged with debris present.
- Two years after the earthquake, only 5% of homes had been rebuilt and many school buildings were still only temporary structures.

Volcanic eruptions in Montserrat

Montserrat is a small island in the Caribbean and was affected by volcanic activity between 1995 and 2013.
- In 1997 a pyroclastic flow killed 19 people.
- The largest settlement, Plymouth, with a population of just 4000, was covered in ash and abandoned.
- Other short-term impacts included evacuation and increased unemployment.

Long-term impacts

- Long-term impacts have included the establishment of an exclusion zone, the creation of the Montserrat Volcano Observatory and the development of new infrastructure and buildings in the north of the island, including homes, hospitals, roads and expansion to the island's port.
- Although there was an economic boom in the early 2000s, once those buildings were built many of the jobs disappeared.
- Thus with fewer jobs in construction, a declining tourist sector and rising prices, many Montserratians left the island for a second time.

Hurricane Matthew, September–October 2016

Hurricane Matthew was a Category 5 hurricane that caused significant loss of life and damage in Haiti, as well as widespread damage in parts of the USA.

Table 3.2.3 Impacts of Hurricane Matthew

Country	Fatalities	Missing	Damage (US$ bn 2016)
Haiti	546	128	$2.8
USA	47	0	$10
Other countries	10	0	$3.7
Total	**603**	**128**	**$16.5 billion**

The most significant impacts were felt in Haiti.
- In the short term there was flooding, high winds, telecommunications were disrupted and damage to over 75% of Jérémie, in the west of the country.
- Hurricane Matthew struck Haiti's south coast. About 175,000 people were made homeless. Around 2.1 million people (20% of Haiti's population) were affected by the hurricane.

Long-term impacts

- Long-term impacts included the redevelopment of the area, clean-up and restoration, and the provision of clean water, sanitation and housing.
- Action Aid and World Nation helped provide clean water, sanitation and shelter.

Revision activity

Try to recall key information about named case studies, e.g. date of event, magnitude, loss of life and any economic losses.

Exam tip

Remember that the loss of life is often much greater in an LIC, whereas the economic loss is often greater in an HIC.

45

+ For those affected by Hurricane Matthew, especially those who lost friends or relatives, the psychological hurt of losing loved ones became a long-term burden.

7 Outline the short-term impacts of the 2015 Nepal earthquake.

8 Briefly explain the long-term problems caused by the eruption of the Soufrière Hills volcano.

9 Compare the impacts of Hurricane Matthew on Haiti and the USA.

3.3 Earthquakes present a hazard to many people and need to be managed carefully

Preparation

There are many ways to prepare for an earthquake including warning and evacuation, building design, remote sensing and geographic information systems (GIS).

The main ways of dealing with earthquakes include:
+ better forecasting and warning
+ building design
+ building location
+ emergency procedures.

There are a number of ways of predicting and monitoring earthquakes. These include:
+ measurement of small-scale ground surface changes
+ ground tilt
+ changes in rock stress
+ clusters of small earthquakes
+ changes in radon gas concentration
+ unusual animal behaviour, especially toads.

Buildings can be designed to cope with the shockwaves that occur in an earthquake. For example, single-storey buildings are more suited than multi-story buildings as the potential for swaying is reduced. Some tall buildings may be built with a 'soft-storey' at the bottom, such as a carpark on raised pillars. This may collapse in an earthquake, so that the upper floors sink down onto it, and this cushions the impact. Building reinforcement strategies include building on foundations built deep into the underlying bedrock, and the use of steel frames that can withstand shaking.

Land-use planning is another way of reducing earthquake risk. Densely populated areas and important services such as hospitals and fire services should not be built close to known fault lines.

Remote sensing such as RapidEye and GeoEye satellite data may be used to monitor changes in ground movement. GIS systems may provide data on land use and infrastructure and highlight areas of particular vulnerability.

> **Geographic information systems (GIS)** Computer systems that allow different types of geographic data to be linked to a location and displayed in an easily understandable form.

> **Revision activity**
>
> Make a list of how it is possible to make buildings more earthquake-proof.

> **Exam tip**
>
> Remember that although it is possible to monitor and predict earthquakes, it is impossible to prevent them.

1 Outline how remote sensing can help predict earthquakes.

2 Explain what is meant by 'land use planning' in relation to earthquakes.

3 Identify ways of predicting and monitoring earthquakes.

Short-term response and relief

Figure 3.3.1 shows a hazard response curve.

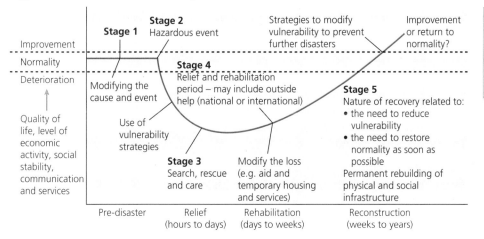

Figure 3.3.1 Hazard response curve

For example, following the Haitian earthquake, the Red Cross estimated that 3 million people needed emergency aid. Seven days after the earthquake, the United Nations had delivered food to only 200,000 people. Assistance – in the form of doctors, trained sniffer dogs, and tents, blankets and food – was pledged from other countries. Financial assistance also poured in. The World Bank led with a $100 million commitment. Yet most of this aid arrived too late for the thousands who were trapped in rubble or waiting for treatment for their injuries.

In contrast, following the 2011 Christchurch (New Zealand) earthquake, in which 185 people died, a full emergency management programme was in place within two hours. Rescue efforts continued for over a week, then shifted to recovery mode.

Revision activity

Make a copy of Figure 3.3.1 and add onto it details about responses following the Nepal earthquake (Section 3.2).

Exam tip

Remember that although the hazard response curve suggests recovery will be complete, in many places, especially in LICs, this does not occur, mainly due to lack of funding.

Now test yourself   TESTED ●

4 Describe the main short-term responses and relief following a large-scale earthquake.

5 Suggest why the short-term responses and relief were more successful in New Zealand than in Haiti.

6 Compare rehabilitation with reconstruction.

Longer-term planning

Risk assessment

Risk is the probability of a hazard event causing harmful consequences (losses in terms of death, injuries, damage to property, the economy and the environment). Most of the risk comes from people living in unsafe housing in areas with known fault lines. However, earthquakes can also occur in areas where no fault line was known to exist.

Earthquakes killed about 1.5 million people in the twentieth century, and the number of people at risk appears to be rising. More than a third of the world's largest and fastest-growing cities are located in regions of high earthquake risk, so the problems are likely to intensify.

Hazard mapping

Hazard mapping shows the most likely areas where earthquakes will occur. Most earthquakes are closely linked with the distribution of fault lines, e.g. in western USA the majority of earthquakes occur in a linear distribution following the San Andreas fault line. However, the timing of earthquakes is difficult to predict.

> **Hazard mapping** Maps that are produced to show the most likely areas that will be impacted by a natural hazard.

Rebuilding programmes

It is difficult to stop an earthquake from happening, so prevention normally involves minimising the prospect of death, injury or damage by controlling building in high-risk areas, and using aseismic designs.

Following earthquakes there is a need for rebuilding programmes. This varies with the scale of the impacts. For example, following the Christchurch earthquakes it was debated whether the whole city would be removed and rebuilt elsewhere. This was never done and the cost would have been far too expensive, even for a relatively rich country.

Rebuilding in poor countries depends largely on the individuals themselves. Following the Haiti (2010) and Nepal (2015) earthquakes, large-scale rebuilding was needed, but the governments were too poor to undertake such measures. Many households are still living in temporary accommodation years after the events.

One option that has been used is to strengthen existing buildings (retro-fitting) to make them safer in an earthquake. Engineers have created a number of 'safe houses' designs which withstand shaking better than some traditional designs. Safe houses can be built cheaply using straw, adobe and old tyres, and by applying a few general principles, e.g. small windows create fewer weak spots in walls. Compressed bales of straw can be sandwiched between layers of plaster to provide some protection from earthquakes.

> **Revision activity**
>
> Define the term 'retro-fitting' and explain how it could make buildings safer during an earthquake.

> **Exam tip**
>
> Remember that although houses can be designed to withstand earthquakes, they may not survive in a high magnitude event nor may they survive secondary hazards such as fires.

> **Now test yourself** TESTED ◯
>
> 7 Describe the variations in earthquake risk in western USA.
> 8 Explain how small windows in a house help in an earthquake.
> 9 Explain how bales of straw can help protect buildings in an earthquake.

Exam practice

1 Identify the correct order for the following four terms. (1 mark)
   A pre-disaster, rehabilitation, relief, reconstruction
   B relief, pre-disaster, reconstruction, rehabilitation
   C pre-disaster, relief, rehabilitation, reconstruction
   D pre-disaster, relief, reconstruction, rehabilitation

2 The population living in which type of area is most at risk following a natural hazard. (1 mark)
   A a rural area
   B an area of low-density buildings
   C a slum
   D an HIC

3 A primary hazard associated with earthquakes is: (1 mark)
   A gas leaks and fires
   B tsunamis
   C ground shaking
   D landslides and rockslides

4 Describe a GIS. (2 marks)

5 Examine the relationship between natural hazard magnitude and frequency. (3 marks)

6 Explain why some people live in hazardous environments. (4 marks)

7 Define the term 'vulnerability'. (1 mark)

8 Describe the main characteristics of the hazard response curve. (4 marks)

9 Examine the causes of hazards associated with earthquakes. (8 marks)

Total: 25 marks

Summary

+ The characteristics, distribution and measurement of tropical cyclones, earthquakes and volcanoes vary.
+ Factors affecting the development of tropical cyclones include ocean temperature, atmospheric (low) pressure, low wind shear and the Coriolis Force.
+ The causes of volcanoes and earthquakes are mainly related to plate boundaries and hot spots.
+ People continue to live in areas at risk from hazards because either they believe the advantages of living there outweigh the disadvantages, or they have no choice.
+ Some countries are more vulnerable than others to the risk of natural hazards because of physical geography (volcanic, by a fault line) and/or in a tropical coastal environment, or they are poor and cannot prepare adequately against natural hazards.
+ There are short-term and long-term impacts of natural hazards.
+ Some countries are more prepared for earthquakes, e.g. warning and evacuation, higher building standards, and use of remote sensing and GIS.
+ Short-term responses and relief following an earthquake focus on emergency aid, shelter and supplies.
+ Long-term planning following an earthquake focuses on rehabilitation and reconstruction, risk assessment and hazard mapping.

3 Hazardous environments

4 Economic activity and energy

4.1 Economic sectors and employment

Classifying production into different economic sectors

The wide range of different jobs that people do can be placed into four economic sectors:

+ The **primary sector** exploits raw materials.
+ The **secondary sector** manufactures primary materials into finished products.
+ The **tertiary sector** provides services.
+ The **quaternary sector** uses high technology to provide information and expertise.

The product chain can be used to show the relationship between the four sectors of employment (Figure 4.1.1).

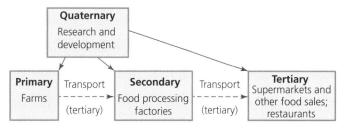

Figure 4.1.1 The food industry's product chain

> **The product chain** The full sequence of activities needed to turn raw materials into a finished product.
>
> **Post-industrial societies** Countries where the tertiary sector dominates employment and where considerable deindustrialisation has occurred.

How employment structure varies and changes over time

As an economy develops, the proportion of people employed in each sector changes (Figure 4.1.2). Countries such as the USA and the UK are 'post-industrial societies' with most people employed in the tertiary sector.

+ In 1900, 40% of employment in the USA was in the primary sector.
+ Increasing mechanisation reduced the demand for workers in primary industries.
+ As these rural jobs disappeared, people moved to urban areas where the secondary and tertiary sectors were expanding.
+ Today only about 2% of employment in the USA is in the primary sector, although production levels are very high due to a very high level of mechanisation and automation.

In manufacturing, robots and other advanced machinery handle assembly-line jobs which once employed large numbers of people. Most manufacturing industries have become more

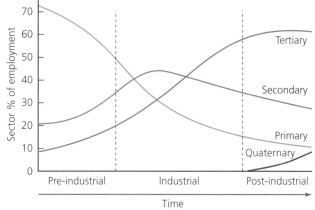

Figure 4.1.2 The Clark-Fisher model

capital intensive. Large processing plants can now be run with a relatively small workforce compared to 30 years ago.

The tertiary sector has also changed as computer networks and other technical advances have reduced the number of people required in some occupations. In developed countries employment in the quaternary sector has become more important – a good measure of how advanced an economy is!

Table 4.1.1 compares the employment structures of three countries. There is a very clear link between employment structure and economic development. The poorest countries of the world are usually primary-product dependent.

> **Capital intensive** A high level of investment in plant and machinery compared to labour.
>
> **Primary-product dependent** When a country relies on one or a small number of primary products for the majority of its export earnings.

Table 4.1.1 Employment structure of a developing country, an emerging country and a developed country

Country	Primary (%)	Secondary (%)	Tertiary (%)
Australia (*developed*)	4	21	75
Malaysia (*emerging*)	11	36	53
Bangladesh (*developing*)	45	30	25

> **Revision activity**
>
> Draw a labelled diagram of the Clark-Fisher model.

Now test yourself   TESTED ◯

1   Which sectors of the economy include a) research scientists, b) farmers, and c) teachers?
2   State the most important economic sector of employment in the pre-industrial stage of the Clark-Fisher model.
3   Define the product chain.

> **Exam tip**
>
> You should take care with the word 'industry' as it can be applied to all sectors of the economy, e.g. the agricultural industry and the service industry.

Factors affecting the location of economic activity

REVISED ◯

Table 4.1.2 Physical and human factors influencing industrial location

Physical factors	Human factors
Site: The availability and cost of land. Large factories will need flat, well drained land on solid bedrock.	**Capital (money)**: Some areas are more likely to attract investment than others.
Raw materials: Industries requiring heavy and bulky raw materials tend to locate as close as possible to these raw materials.	**Labour**: The quality and cost of labour are most important. The reputation, turnover, mobility and quantity of labour can also be important.
Energy: Energy-hungry industries, e.g. metal smelting, may be drawn to countries with relatively cheap hydroelectricity, e.g. Norway.	**Transport and communications**: Transport costs remain important for heavy, bulky items. Accessibility to airports, ports, motorways etc. may be crucial for some industries.
Natural routeways and harbours: Many modern roads and railways still follow natural routeways. Natural harbours provide good locations for ports and the industrial complexes often found at ports.	**Markets**: The location and size of markets is a major influence for some industries.
	Government influence: Government policies and decisions can have a big direct and indirect impact on the location of industry.
Climate: Some industries such as aerospace and film benefit directly from a sunny climate. Indirect benefits include lower heating bills and a more favourable quality of life.	**Quality of life**: Highly skilled personnel will favour areas where the quality of life is high.

Table 4.1.2 shows the factors affecting the location of manufacturing industry. These factors can be applied to varying degrees to the other sectors of employment. Changes in the importance of any of these factors can have a big effect on employment in regions and countries.

For example:

Primary sector

+ Investment in new irrigation projects can extend an area under farming.
+ The exhaustion of mineral deposits can shut entire industries such as coal mining.

Tertiary sector

+ Improving the accessibility of a central business district (CBD) may attract new businesses.
+ Government grants to encourage tourism can help create new jobs.

Quaternary sector

+ Higher national investment in science and technology can attract new high-tech industries.
+ Improvements in the quality of life in a region will make it more attractive to high tech firms.

Increasingly, manufacturing and large retail companies have located together on industrial estates or business/retail parks. These are usually located close to good quality transport infrastructure, often in outer suburban or urban fringe locations. The movement of manufacturing and retail outwards from CBDs and inner cities is known as decentralisation.

The changing location of manufacturing

Changes in the location of manufacturing industry have occurred at a range of scales:

+ The global shift from the developed world to emerging and developing countries.
+ The movement from traditional manufacturing areas to higher quality of life regions in developed countries.
+ A regional movement from urban areas towards 'greenfield' rural locations.

> **Now test yourself** TESTED ⬤
>
> 4   State three human factors affecting the location of industry.
> 5   Define decentralisation.
> 6   What is a 'greenfield' location?

> **Industrial estate** An area zoned and planned for the purpose of industrial development. Manufacturing industry has a significant presence.
>
> **Business/retail parks** Similar to industrial estates, but dominated by tertiary activities.
>
> **Decentralisation** The outward movement of economic activity from CBDs and inner cities.
>
> **Greenfield locations** Areas of agricultural land or other undeveloped sites earmarked for construction.

> **Revision activity**
>
> What are the advantages of companies locating together on industrial estates?

> **Exam tip**
>
> Remember that many geographical changes can occur at a variety of scales, from local to global.

Reasons for change in economic activity REVISED ⬤

Raw materials

+ Farm production can be decimated due to drought, soil erosion and insect plagues.
+ The exhaustion of a raw material deposit can cause job losses, not only in mining, but also in local industries using the raw material. Other businesses will suffer as jobs are lost and incomes fall.
+ Technological advances can reduce the amount of raw materials required or change the balance between the different raw materials used. Such changes can affect location decision making.

Globalisation

+ Globalisation is a term that describes increasing global links (economic, cultural and political).
+ Until the post-1950 period, industrial production was mainly organised within individual countries. Since then, production has been increasingly divided into different skills that are often spread across a number of countries.
+ Transnational corporations (TNCs) have increasingly been able to take advantage of lower costs in developing and emerging countries.

> **Globalisation** The increasing interconnectedness and interdependence of the world economically, culturally and politically.
>
> **Transnational corporations** Firms that own or control productive operations in more than one country through foreign direct investment (FDI).

- The result is that many of the traditional industries of developed countries such as iron and steel, shipbuilding and textiles have moved on a large scale to emerging economies such as China, India and Brazil.
- The large-scale loss of manufacturing employment in developed countries is known as deindustrialisation.

The role of technology

Advances in technology have affected all sectors of employment. Large farms can now be run by a small workforce due to advanced mechanisation and automation. The internet has allowed large companies to manage complex operations all over the world. Many jobs in service industries once located in developed countries have been outsourced to emerging economies.

Demographic change

Global population has risen from 3 billion in 1960 to 7.8 billion today. Such a large population change has led to a great increase in the demand for goods and services worldwide. At the same time, a growing population has resulted in a larger supply of potential workers. The rate of population change varies from country to country. Other factors to consider include:
- The age structure of a population affecting the size of the working population and the demand for particular goods and services.
- Average disposable income and income distribution.
- Cultural traits affecting demand for different goods and services.

Government policy

Governments influence industrial location for economic, social and political reasons. Government at national and regional levels may use grants to attract major international companies. There is a high level of competition both between and within countries to attract inward investment. Examples of major government spending policies include building new high-speed railways and airport expansion. Levels of employment in different sectors can be affected by such decisions.

Now test yourself TESTED ◯

7 What is globalisation?
8 Suggest two ways in which changes to the age structure of a population can affect economic activity.
9 Give an example of how government investment can influence the location of economic activity.

Emerging economies
National economies which are becoming more advanced, usually through rapid growth and industrialisation.

Deindustrialisation
The long-term decline in employment in manufacturing industry.

Outsourcing When a company contracts out work to another company. Such agreements are often between companies in high-wage developed countries and lower-wage emerging countries.

Revision activity

For an industry you have studied, list the raw materials required.

Exam tip

Remember that globalisation has had advantages and disadvantages at all scales, from the local to the global. It is a process of change that has consequences.

4.2 The growth and decline of different sectors – impacts and resource issues

Positive and negative impacts of change in economic sectors

REVISED ◯

Sector shift in the UK

The UK is an example of a developed country in the 'post-industrial' stage of the Clark-Fisher model. The Industrial Revolution began in the UK in the late eighteenth century and then spread to other countries such as Germany and the USA. The number of people working in services in the UK overtook the manufacturing workforce in 1881, over 130 years before this happened in China. Over 80% of the UK workforce is now in the tertiary sector. No region in the country has less than 70% and London has 91%.

+ The City of London contains one of the greatest concentrations of high-level tertiary industry in the world.
+ It is one of the big three financial centres, along with New York and Tokyo.
+ Among the important buildings in the City of London are the Bank of England, the London Stock Exchange and Lloyd's of London (insurance).

Table 4.2.1 summarises some of the positive and negative aspects of sector shift in the UK.

Table 4.2.1 Examples of positive and negative impacts of sector change

Positive	Negative
The UK attracts a very high level of foreign direct investment because of its expertise in a wide range of secondary and tertiary industries.	High job losses in traditional secondary industries such as coal, iron and steel, shipbuilding and textiles due to deindustrialisation.
The UK has become one of the world's leading exporters of tertiary products, which creates considerable wealth for the country.	A cycle of deprivation in inner cities and other areas affected by large-scale manufacturing decline.
Low-cost manufactured goods from China and other emerging countries have helped keep inflation low.	A widening gap between a) the highest and lowest paid workers and b) the richest and poorest regions.
Deindustrialisation, with subsequent landscape renewal, has improved environmental conditions in many parts of the UK.	Transnational corporations can move investment away from a country as quickly as they can bring it in, causing loss of jobs and corporation tax.

Sector shift in China

China is an example of an emerging country. Its economy has grown rapidly in the last forty years. Before this time China was largely an agricultural economy. More recently the secondary and tertiary sectors have grown at a very fast rate. China is in the 'industrial' stage of Clark-Fisher model.

Major economic reforms introduced in 1978 aimed to:
+ rapidly develop manufacturing industry
+ extend China's global trade links
+ increase the rate of economic growth.

Over 70% of the workforce was employed in agriculture in 1978. By 2018, it was down to 27%. The tertiary sector overtook the secondary sector in 2013. Retailing and tourism are examples of fast-growing tertiary activities in China. China is a) the world's largest manufacturing economy, and b) the world's largest exporter of goods. However, average incomes are still well below developed countries such as the UK.

Table 4.2.2 Sector change in China (GDP)

Sector	Year 2000 (%)	Year 2017 (%)
Primary	14.7	7.9
Secondary	45.5	40.5
Tertiary	39.8	51.6

City of London The historic centre of London containing the main financial district. It is just over one square mile in area.

Sector shift The changing distribution of employment between economic sectors.

Economic reforms New government policies designed to improve the economic performance of a country. Economic reforms can stimulate sector shift.

Major changes in a country's economy can bring many benefits, but problems can also develop (Table 4.2.3).

Table 4.2.3 Examples of positive and negative impacts of sector change

Positive	Negative
A large increase in average wages and living standards.	An increasing gap between 'richer' urban and 'poorer' rural living standards.
Much improved 'soft' infrastructure – housing, health, education.	Wider regional imbalance between the fastest and slowest growing regions.
More highly developed 'hard' infrastructure – roads, railways, airports, energy supply networks.	Very high levels of pollution in large urban-industrial areas.
Increasing levels of foreign investment.	An increasing contribution to global climate change. Increasing production of CO_2.

Revision activity

Draw a simple graph to illustrate the data in Table 4.2.2.

Now test yourself   TESTED ◯

1 Where is the UK's biggest concentration of high-level tertiary industry?
2 Name two traditional industries that have declined in the UK due to deindustrialisation.
3 List China's economic sectors by order of size in a) 2000 and b) 2017.

Exam tip

Emerging countries are also referred to as newly industrialised countries in some textbooks.

Case study

Informal employment in a megacity: Mumbai, India

Over two billion people worldwide work in the informal sector. The largest concentrations of informal sector work are the megacities of developing and emerging countries. Here, levels of unemployment and underemployment are often very high.

About 68% of Mumbai's workforce is estimated to work in the informal sector. The population of this huge urban area has grown rapidly. With intense competition for formal sector jobs, most people have to look to the informal sector to meet their basic needs. The characteristics of the informal sector are:
+ highly labour intensive
+ low and unreliable pay
+ work often temporary and/or part-time
+ poor job security, with an absence of fringe benefits
+ poor working conditions and a high exposure to health and safety risks
+ high potential for exploitation, especially for children.

The four leading informal sector occupations in Mumbai are domestic workers, home-based workers, street vendors and waste pickers. Other occupations include messengers and repair-shop workers. Over three-quarters of Mumbai's informal workers are in the service sector. Informal manufacturing includes both the workshop sector and the traditional craft sector. A big benefit of the sector is that it provides cheap goods and services for the lower income population in the city.

An often-stated disadvantage of the informal sector to the economy is that it operates outside the tax system. However, most informal workers have incomes below the threshold for paying tax. Most of Mumbai's informal workers live in slums. Dharavi is not only Mumbai's largest slum, it is also the biggest in Asia.

India wants to gradually 'transition' informal workers into the formal sector. As this involves such a large number of people, it will be a massive challenge!

Informal sector The part of the economy operating outside official government recognition.

Megacity A city with a population over 10 million.

Underemployment When people are working less than they would like to in order to earn a reasonable living.

Formal sector The regulated part of the economy in terms of taxation and other government rules and regulations.

Slum An area of poor quality housing, usually densely populated, in a poor state of repair, and with inadequate services.

Revision activity

List the four leading informal sector occupations in Mumbai.

Now test yourself   TESTED ◯

4 Why is Mumbai described as a megacity?
5 Define the informal sector.
6 List four characteristics of the informal sector.

Exam tip

It is important to be clear about the distinction between unemployment and underemployment.

Theories used to explain the relationship between population and resources

As a country develops, the highest average living standards mark the optimum population in economic terms (Figure 4.2.1). Before that population is reached, a country could be said to be underpopulated. As the population rises beyond the optimum, a country can be said to be overpopulated.

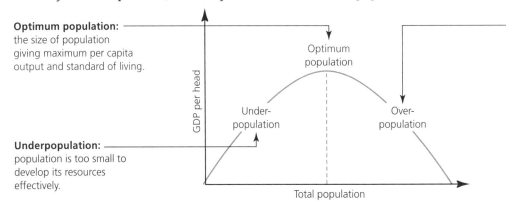

Optimum population: the size of population giving maximum per capita output and standard of living.

Underpopulation: population is too small to develop its resources effectively.

Overpopulation: an increase in population or decrease in natural resources which leads to a decrease in standards of living for the population as a whole. This is related to the *carrying capacity* of a country.

Figure 4.2.1 The optimum population

There are many indications that the human population is pushing up against the limits of the Earth's resources. For example:
+ One-quarter of the world's children have protein-energy malnutrition.
+ Water scarcity affects every continent.

The most obvious examples of population pressure are in the developing world.

The ideas of Thomas Malthus

The Reverend Thomas Malthus produced his *Essay on the Principle of Population* in 1798. He took a pessimistic view about the relationship between population and resources.
+ He maintained that while the supply of food could at best be increased by a constant amount in arithmetical progression (2 – 3 – 4 – 5 – 6), the human population tends to increase in geometrical progression (2 – 4 – 8 – 16 – 32).
+ In time, population would outstrip food supply until a catastrophe occurred in the form of famine, disease or war.
+ These limiting factors maintained a balance between population and resources in the long term.

Malthus could not have foreseen the great advances that were to unfold in the following two centuries. However, today nearly all of the world's productive land is already exploited. Most of the unexploited land is either too steep, too wet, too dry or too cold for agriculture. Modern-day resource pessimists are known as neo-Malthusians. In recent years neo-Malthusians have highlighted:
+ the steady global decline in the area of farmland per person
+ the steep rise in the cost of many food products in recent years
+ the growing scarcity of fish in many parts of the world
+ the continuing increase in the world's population.

The resource optimists, such as Danish economist Ester Boserup, believe that human ingenuity will continue to conquer resource problems. They point to:
+ the development of new resources
+ the replacement of less efficient with more efficient resources
+ the rapid development of green technology, with increasing research and development in this growing economic sector.

Optimum population The population that achieves a given aim in the most satisfactory way. Often viewed as the population resulting in the highest standard of living.

Overpopulated When there are too many people relative to the resources and the level of technology available.

Underpopulated When there are too few people in an area to use the resources available efficiently.

Population pressure When population per unit area exceeds the carrying capacity.

Exam tip

Figure 4.2.1 shows the optimum population in economic terms. The figure at which this is achieved might be different if the optimum was seen in social or environmental terms.

Revision activity

Create a table to show countries that might be considered to be a) underpopulated, and b) overpopulated.

4.3 Energy resource management

Energy demand and resource management

REVISED ⭕

Energy demand and production

+ Global energy demand increased by over 60% between 1993 and 2018.
+ Fossil fuels dominate the global energy mix. Their relative contribution to primary energy consumption in 2018 was: oil – 33.6%, coal – 27.2%, natural gas – 23.9%.
+ In contrast, hydroelectricity accounted for 6.8%, nuclear energy 4.4% and renewable energy 4.1%.

> **Energy mix** The relative contribution of different energy sources to a country's energy consumption.
>
> **Primary energy** An energy form found in nature that has not been changed by human processing, e.g. fossil fuels to electricity.

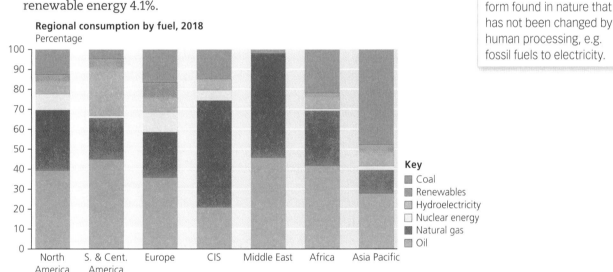

Regional consumption by fuel, 2018
Percentage

Key
- Coal
- Renewables
- Hydroelectricity
- Nuclear energy
- Natural gas
- Oil

Figure 4.3.1 Regional consumption of fuel by type

National demand is mainly down to:
+ the size of a country's population and its rate of growth
+ its level of economic development, particularly wealth and level of technology.

Table 4.3.1 shows per capita energy consumption by world region in 2018. Average consumption in North America is 16 times higher than in Africa.

Table 4.3.1 Primary energy consumption per capita, 2018

World region	Per capita energy consumption (GJ), 2018
North America	239.8
South and Central America	56.4
Europe	127.4
CIS	160.9
Middle East	148.5
Africa	15.0
Asia Pacific	60.2
World	76.0

Figure 4.3.2 is a model showing the relationship between resource use in general and the level of economic development. This model applies well to energy consumption. Growth in energy demand is particularly rapid in emerging economies such as China, India and Brazil. The key factor in energy supply is energy resource endowment. However, resources by themselves do not constitute supply. Capital and technology are required to exploit energy resources. In developing countries, about 2.5 billion people rely on fuelwood as their main source of energy. The transition from fuelwood and animal dung to 'higher level' sources of energy, the energy ladder, occurs as part of the process of economic development.

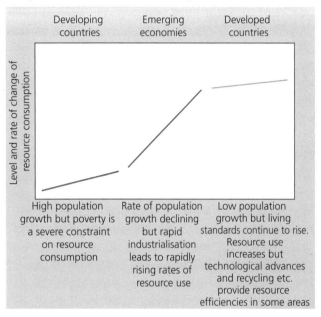

Figure 4.3.2 Model of the relationship between resource use and the level of economic development

The energy gap and energy security

An increasing number of countries are facing an 'energy gap'. This is the difference between a country's rising demand for energy and its ability to satisfy this demand from its own resources. The International Energy Agency (IEA) defines energy security as the 'uninterrupted availability of energy sources at an affordable price'. Energy security can be affected by a number of factors. Long-term energy security requires careful planning and considerable investment in line with economic development and energy needs.

Energy supply interruption can be caused by a number of factors including:
+ natural hazards
+ civil disturbance
+ geopolitics
+ rapid resource depletion
+ affordability
+ environmental concerns.

> **Energy ladder** The transition from fuelwood and animal dung to 'higher-level' sources of energy such as electricity.
>
> **Energy gap** The difference between a country's demand for energy and its ability to satisfy this demand from its own resources.
>
> **Energy security** The uninterrupted availability of energy sources at an affordable price.

> **Revision activity**
>
> List the sources of energy in order of their contribution to the global energy mix.

> **Exam tip**
>
> The energy ladder is a part of the development process. It requires investment linked to rising standards of living.

> **Now test yourself** TESTED ◯
>
> 1 Which country is the largest total consumer of energy?
> 2 Which world regions have the highest and lowest energy consumption per capita?
> 3 List three factors that can cause an energy supply interruption.

Non-renewable and renewable energy resources

REVISED

Non-renewable sources of energy are the fossil fuels and nuclear fuel. Eventually, these resources could become completely exhausted. Fossil fuels are the major source of greenhouse gas emissions. Climate change due to these emissions is the biggest environmental problem facing the planet.

Renewable energy resources cause little or no pollution. Renewable energy includes hydroelectricity, biofuels, wind, solar, geothermal, tidal and wave power. Hydroelectricity is the one renewable source of energy that is sometimes described as a traditional source of energy because water power has been used to generate electricity for over one hundred years.

Non-renewable resources still dominate global energy supply. The challenge is to transform the global energy mix to achieve a better balance between renewables and non-renewables.

Countries are eager to harness renewable energy resources to:
+ reduce their reliance on domestic fossil fuel resources
+ lower their reliance on costly fossil fuel imports
+ improve their energy security
+ cut greenhouse gas emissions.

The cost gap with non-renewable energy has narrowed and even closed in some cases. Figure 4.3.3 shows the sharp increase in the consumption of renewable energy (other than HEP) in the last decade. In 2015, this accounted for almost 2.8% of global primary energy consumption. The newer sources of renewable energy making the largest contribution to global energy supply are wind power and biofuels.

Non-renewable energy Types of energy that cannot be replaced after they have been used.

Fossil fuels Fuels consisting of hydrocarbons (coal, oil and natural gas).

Renewable energy Sources of energy such as solar and wind power that are not depleted as they are used.

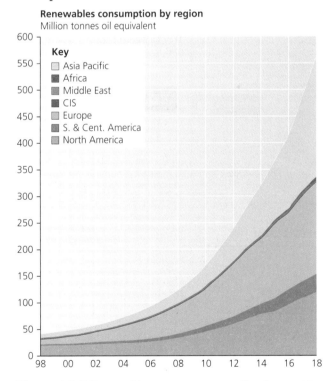

Renewables consumption by region
Million tonnes oil equivalent

Key
- Asia Pacific
- Africa
- Middle East
- CIS
- Europe
- S. & Cent. America
- North America

Figure 4.3.3 Renewable energy consumption by world region 1998–2018

Energy sources: Advantages and disadvantages

All sources of energy have advantages and disadvantages. The important issue for each energy source is the balance between the positive and negative factors. Two sources of energy are considered here – nuclear and solar. You should be able to produce similar lists or tables for all the other sources of energy as part of your revision programme.

Nuclear power Electricity generated by a nuclear reactor. Energy is created by splitting apart the nuclei of atoms.

4 Economic activity and energy

59

Disadvantages of nuclear power

+ Power plant accidents, which could release radiation into air, land and sea.
+ Radioactive waste storage/disposal. Most concern is over the small proportion of 'high-level waste'.
+ Rogue state or terrorist use of nuclear fuel for weapons.
+ High construction and decommissioning costs.
+ The possible increase in certain types of cancer near nuclear plants.

Advantages of nuclear power

+ Zero emissions of greenhouse gases.
+ Reduced reliance on imported fossil fuels.
+ Not as vulnerable to fuel price fluctuations as oil and gas.
+ In recent years nuclear plants have demonstrated a very high level of reliability.
+ Nuclear technology has spin-offs in fields such as medicine and agriculture.

Solar power

Table 4.3.2 lists the advantages and disadvantages of solar power.

Table 4.3.2 Advantages and disadvantages of solar power

Advantages	Disadvantages
A completely renewable resource	Initial high cost of solar plants
No noise or direct pollution	Solar power cannot be harnessed during storms, on cloudy days or at night
Very limited maintenance required	Of limited use in countries with low annual hours of sunshine
Technology is improving and reducing costs	Large areas of land required to capture the sun's energy in order to generate significant amounts of power
Can be used in remote areas where it is too expensive to extend the electricity grid	
A generally positive public perception	

> **Revision activity**
>
> Make lists of energy sources which are renewable and non-renewable.

Now test yourself

TESTED ◯

4 What is renewable energy?
5 Define energy mix.
6 State three reasons why countries are keen to increase their renewable energy production.

> **Exam tip**
>
> Solar power is generally taken to mean the production of solar electricity, as distinct from solar hot water systems.

Managing energy resources sustainably

REVISED ◯

Energy efficiency

Meeting future energy needs while avoiding serious environmental degradation will require increased emphasis on education, efficiency and conservation in all sectors of the economy.

Policies have been developed which include:

+ much greater investment in renewable energy
+ conservation
+ recycling
+ carbon credits
+ 'green' taxation.

Managing energy supply is often about balancing socio-economic and environmental needs. Many countries are looking increasingly at the concept of community energy. Much energy is lost in transmission if the source of supply is a long way away. Energy produced locally is much more efficient. This will invariably involve microgeneration.

> **Community energy** Energy produced close to the point of consumption.
>
> **Microgeneration** Refers to generators producing electricity with an output of less than 50 KW.

Table 4.3.3 summarises some of the measures governments and individuals can undertake to reduce the demand for energy and thus move towards a more sustainable situation.

Table 4.3.3 Examples of energy conservation measures

Government	Individuals
+ Improve public transport. + Set a high level of tax on petrol. + Set minimum fuel consumption requirements for vehicles. + Congestion charging to deter non-essential car use in city centres. + Encourage business to monitor and reduce its energy usage. + Promote investment in renewable forms of energy. + Pass laws to compel manufacturers to produce higher efficiency electrical products.	+ Walk rather than drive for short local journeys. + Buy low fuel consumption/low emission cars. + Reduce car usage by planning more multi-purpose trips. + Use public rather than private transport. + Car pooling. + Use low-energy light bulbs. + Install and improve home insulation. + Turn boiler and radiator settings down. + Wash clothes at lower temperatures. + Purchase energy-efficient appliances.

Revision activity

What examples of microgeneration are evident in the region in which you live?

Exam tip

It is easy to think that governments could do much more to conserve energy, but many measures affect individual and family budgets. Governments need to convince people that higher costs are justified.

Now test yourself

TESTED ◯

7 Define community energy.

8 List four energy conservation measures that individual people can adopt.

Case study

Energy resource management in China

China uses more energy than any other country in the world. In 2018, China's main sources of energy were: coal (58.2%), oil (19.6%) and hydroelectricity (8.3%). Chinese investment in energy resources abroad has risen rapidly in order to achieve long-term energy security. In recent years, China has tried to take a more balanced approach to energy supply and to reduce its environmental impact:

+ The development of clean coal technology is an important aspect of this approach.
+ The further development of nuclear and hydropower is another important strand of Chinese policy.
+ China aims to increase the production of oil while augmenting that of natural gas and improving the national oil and gas network. Priority was also given to building up the national oil reserve.
+ Total renewable energy capacity in China reached 502 GW in 2015. This included 319 GW of hydroelectricity, 129 GW of wind energy, 43 GW of solar PV and 10 GW of bioenergy.
+ The Three Gorges Dam across the Yangtze River is the world's largest electricity generating plant of any kind. This is a major part of China's policy in reducing its reliance on coal.

China's wind power capacity has also grown rapidly in recent years. It is now the largest in the world by a large margin. China plans to have a total renewable capacity of more than 800 GW by 2021 (Table 4.3.4). In 2018, China accounted for 45% of the global growth in renewable power generation. It is now the largest producer of renewable power in the world. However, only 4.4% of China's primary energy production comes from renewables.

Table 4.3.4 China's projected renewable electricity capacity (GW), 2015–21

	2015	2018	2021
Hydropower	319.4	348.4	368.4
Bioenergy	10.3	14.4	18.4
Wind:	129.3	189.4	257.1
Onshore	128.3	186.8	250.3
Offshore	1.0	2.6	6.8
Solar PV	43.2	104.2	160.2
CSP/STE	0.0	1.5	3.1
Geothermal	0.0	0.1	0.1
Ocean	0.0	0.0	0.0
Total	**502.3**	**557.8**	**807.3**

Now test yourself

TESTED ◯

9 What is 'clean coal technology'?

10 Describe the projected increase in renewable electricity capacity.

Case study

Energy resource management in Sweden

Few countries consume more energy per capita than Sweden, but Swedish carbon emissions are very low. In 2018, Sweden's primary consumption energy mix was: oil – 27.6%; natural gas – 1.3%; coal – 3.7%; nuclear – 28.9%; hydro – 26.1%; renewables – 12.3%.

+ Hydropower and bioenergy are the top renewable resources. Sweden has many fast-flowing rivers and forests cover 63% of its land area.
+ About 80% of electricity production comes from nuclear energy and hydropower. However, nuclear power is a controversial subject in Sweden.
+ About 11% of electricity production is from wind power. This is set to increase sharply in the future.
+ Combined heat and power (CHP) plants account for 9% of electrical output. These plants are mainly powered by biofuels.

+ Solar power remains limited in capacity largely due to Sweden's northerly latitude, but it is growing with the aid of government funding.
+ The number of heat pumps has increased sharply since the 1990s. These pumps use renewable energy by transferring heat, mainly from the ground. This reduces demand on electricity from the grid.
+ Carbon taxation has proved effective in energy change and efficiency.

The long-term aims of Sweden's energy policy are:
+ a transition to 100% renewable electricity production by 2040
+ a zero-carbon economy by 2045.

Sweden plans to achieve 50% more efficient energy use by 2030. Since the 1980s, Sweden's population has increased by 25% and GDP has doubled, but Sweden is now using less energy and electricity than then. Sweden also plans to be at the forefront of energy storage.

Now test yourself

11 Which two sources of energy dominate electricity production in Sweden?

12 List the long-term aims of Sweden's energy policy.

Exam practice

1 Which of the following is in the primary sector: banking, biotechnology, farming, teaching? (1 mark)

2 a) Name a graphical method that can be used to compare the employment structures of a large number of countries. (1 mark)

   b) Which sector of employment is most important in post-industrial countries? (1 mark)

   c) The poorest countries of the world are often 'primary product dependent'. What does this mean? (2 marks)

3 How can the geographical location of raw materials influence where a factory is sited? (2 marks)

4 State three characteristics of the informal sector of employment. (3 marks)

5 Define energy security. (2 marks)

6 Discuss the advantages of one source of renewable energy. (5 marks)

7 Examine energy resource management in a named country. (8 marks)

Total: 25 marks

Summary

+ The wide range of different jobs that people do can be placed into four economic sectors: primary, secondary, tertiary and quaternary.
+ The Clark-Fisher model shows how employment structure varies at different levels of economic development.
+ A range of human and physical factors can affect the location of economic activity and these factors can change over time.
+ The reasons for changes in the number of people employed in each economic sector include the availability of raw materials, technological advances, globalisation, demographic changes and government policies.
+ Economic sector shifts can create both positive and negative impacts.
+ The largest concentrations of informal sector workers are in the megacities of developing and emerging countries.

+ The different theories of Malthus and Boserup can be used to explain the relationship between population and resources.
+ Energy demand and production vary globally and are affected by factors including population growth, increased wealth and technological advances.
+ Both non-renewable and renewable sources of energy have positive and negative impacts on people and the environment.
+ An increasing number of countries are facing an energy gap.
+ Energy security can be affected by a number of factors. Long-term energy security requires careful planning and considerable investment in line with economic development and energy needs.
+ Meeting future energy needs while avoiding serious environmental degradation will require increased emphasis on education, efficiency and conservation in all sectors of the economy.

5 Rural environments

5.1 Rural environments are natural ecosystems exploited by human activities

Distribution and characteristics of the world's biomes

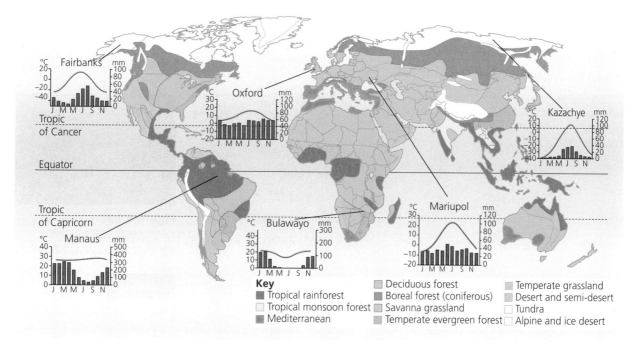

Figure 5.1.1 Distribution and characteristics of the world's biomes

Tropical rainforests are generally located in low latitude areas with rainfall of over 1700 mm and annual temperatures of over 26–27 °C. Seasonal range of temperature is low and the growing season is year-round. Vegetation is evergreen allowing photosynthesis to occur year-round. The forests are layered, very diverse and very productive. Soils are very infertile.

Temperate grasslands are mainly found in temperate continental areas. They have a continental climate, with hot summers and cold winters. Precipitation is low, with snow in winter and convectional rain in summer. Vegetation is mainly grass with occasional trees near rivers

Biome A collection of ecosystems sharing similar climatic conditions.

Revision activity

Write a description of all of the climate graphs shown in Figure 5.1.1. Make sure you give each one a title.

Exam tip

When describing climate patterns, refer to temperature (maximum and minimum) and seasonal variations, and rainfall total and seasonality.

Now test yourself

TESTED

1 Describe the climate of Kazachye, a **tundra** area.
2 Describe the climatic conditions in Bulawayo, an area of tropical grassland (savanna).
3 Describe the main characteristics of the climate associated with deciduous forests.

Ecosystem goods and services

There are four main types of ecosystem service:

+ **Supporting services** are the essentials for life and include primary productivity, soil formation and the cycling of nutrients. These support all other types of ecosystem goods and services.
+ **Regulating services** are a diverse set of services and include pollination, regulation of pests and diseases and production of goods such as food, fibres and wood. Other services include climate and hazard regulation and water quality regulation.
+ **Provisioning services** are the services people obtain from ecosystems such as food, fibre, fuel (peat, wood and non-woody biomass) and water from aquifers, rivers and lakes. Goods can be from heavily managed ecosystems (intensive farms and fish farms) or from semi-natural ones (such as by hunting and fishing).
+ **Cultural services** are derived from places where people's interaction with nature provides cultural goods and benefits. Open spaces – such as gardens, parks, rivers, forests, lakes, the seashore and wilderness areas – provide opportunity for outdoor recreation, learning, spiritual well-being and improvements to human health.

Boreal forest Coniferous forests found in high latitudes of the northern hemisphere.

Tundra Grassland, lichen and moss, with occasional dwarf trees, associated with relatively cold, high latitude environments.

Ecosystem goods The products of ecosystems which can be sold or used for sustenance by people.

Ecosystem services The services that ecosystems provide, e.g. climate and flood regulation.

Table 5.1.1 Ecosystem services

Services	Mountains, moorlands and heaths	Woodlands
Provisioning services (goods)	Food*	Timber*
	Fibre*	Species diversity*
	Fuel*	Fuelwood*
	Fresh water*	Fresh water*
Regulating services	Climate regulation†	Climate regulation†
	Flood regulation†	Flood regulation†
	Wildfire regulation†	Erosion control†
	Water quality regulation†	Disease and pest control†
	Erosion control†	Wildfire regulation†
		Air and water quality regulation†
		Soil quality regulation†
		Noise regulation†
Cultural services	Recreation and tourism*	Recreation and tourism*
	Aesthetic values*	Aesthetic values*
	Cultural heritage*	Cultural heritage*
	Spiritual values*	Employment*
	Education*	Education*
	Sense of place*	Sense of place*
	Health benefits*	Health benefits*

Key: *Goods / †Services

Figure 5.1.2 Tropical rain forests provide many ecological goods and services

Exam tip

Make a list of the ecosystem services provided by a local area close to your school/home area.

Now test yourself

TESTED

4 Distinguish between provisioning services (goods) and regulating services.
5 Explain how forests help regulate climate.
6 Explain the importance of two cultural services.

Answers to Now Test Yourself and Exam Practice questions at **www.hoddereducation.co.uk/myrevisionnotesdownload**

Human modification of ecosystems: farming systems

People use, modify and change ecosystems and rural environments to obtain food through farming systems. Farming systems can be divided into four main pairs (Table 5.1.2). These are not exclusive categories but indicate a scale, along which all farming types can be placed.

> **Arable** Farming systems associated with crops.
>
> **Pastoral** Farming systems associated with livestock.

Table 5.1.2 The world's major farming systems

Arable: the cultivation of crops such as wheat farming in the Great Plains of the USA
Pastoral: rearing animals, e.g. sheep farming in New Zealand
Commercial: products are sold to make a profit such as market gardening in the Netherlands
Subsistence (or peasant): products are consumed by the cultivators as in the case of shifting cultivation by the Kayapo in the Amazonian rainforest
Intensive: high inputs or yields per unit area such as battery-hen production
Extensive: low inputs or yields per unit area as in free-range chicken production
Nomadic: farmers move seasonally with their herds such as the Pokot, pastoralists in Kenya
Sedentary: farmers remain in the same place throughout the year, e.g. dairy farming in Devon and Cornwall

A farming system is a simplified way of showing the factors that influence farming, the processes that occur on the farm, and the products produced (Table 5.1.3).

> **Commercial** Farming systems in which the aim is to sell goods and make a profit.
>
> **Subsistence** Farming systems which largely consume all the products on the farm/homestead.
>
> **Intensive** Large amounts of inputs/outputs per unit area.
>
> **Extensive** Small amounts of inputs/outputs per unit area.
>
> **Biodiversity** Biological diversity which includes habitat diversity, species diversity and genetic diversity (within a species).

Table 5.1.3 A systems diagram for a wheat farm

Inputs	Processes	Outputs
Land	Preparing land	Main product (wheat grain for sale for further processing)
Energy	Ploughing	
Labour	Harrowing	By-product (straw bales for animal feed)
Machinery	Manuring	
Government subsidies	Sowing	Waste products (stubble, burned or ploughed in to enrich next year's soil)
Fertilisers	Fertilising	
Pesticides	Weeding	
Seeds	Pest control	
	Harvesting	

Farming systems have a major impact on natural ecosystems. They simplify them and are usually associated with a limited number of types of crops or livestock (Table 5.1.4).

> **Revision activity**
>
> Make a systems diagram/table for two contrasting types of farming in your home country.

Table 5.1.4 A comparison of natural and agricultural ecosystems

	Natural	Agricultural
Food web	Complex – several layers	Simple – mostly one or two layers
Biomass	Large – mixed plant and animal	Small – mostly plant
Biodiversity	High	Low – often monoculture
Gene pool	High	Low, e.g. three species of cotton amount for 53% of crop
Nutrient cycling	Slow, self-contained, unaffected by external supplies	Largely supported by external supplies
Productivity	High	Lower
Modification	Limited	Extensive – inputs of feed, seed, water, fertilisers, energy fuel, outputs of products, waste, etc.

Now test yourself TESTED ◯

7 Outline the main changes to ecosystems as a result of agriculture.
8 Explain the term monoculture.
9 Explain how agriculture can be considered to operate as a system.

Exam tip

You should be able to describe any farming types in terms of the four pairs of definitions in Table 5.1.2.

5.2 Rural environments have contrasting physical, social and economic characteristics and are experiencing significant changes

Characteristics of a rural environment

REVISED ◯

Rural areas are very varied. Nevertheless, they can be described in terms of their landscape, climate, settlement, population, land use, employment, accessibility and management (development or conservation).

Rural landscapes influence what activities may take place. Steep slopes with thin, infertile soils limit the potential for farming.

The climate also influences the potential for different activities. Areas that are too hot/cold, wet/dry may be more limited in their economic activities. Areas with low population densities may develop extensive farming, whereas areas with high population density may develop more intensive forms of farming.

Figure 5.2.1 Rural landscape, west coast of Ireland

Much of the settlement in rural areas is of small villages and hamlets. Population densities are generally low, although some river valleys may have quite high densities, e.g. the Ganges and the Nile valleys. In many rural areas, due to the lack of well-paid employment opportunities, younger people migrate to urban areas. Hence, rural populations in many places have ageing populations.

Land use in most rural areas is related to primary activities, especially farming. In many HICs, some workers may live in rural areas but work from home (telecommuting). Increasingly, leisure and tourism are becoming important.

Rural areas generally have poorer accessibility than urban areas. However, rural areas that are close to large urban areas have better accessibility than those further away, and are experiencing more change. Management of rural areas may involve conservation, e.g. national parks and country parks, while some rural areas may be developed, e.g. irrigation, housing and leisure developments.

Telecommuting People who work from home through the use of computers/the internet.

Revision activity

For a rural area you have studied, describe its main characteristics.

Exam tip

Remember that there is considerable variation in rural areas but the contrasts between rural and urban are greater.

Now test yourself TESTED ◯

1 Describe the landscape in Figure 5.2.1.
2 Identify the likely farming type to occur in the landscape shown in Figure 5.2.1.
3 Describe the main land use/activity associated with most rural areas.

Rural change in developed countries

Figure 5.2.2 shows a model of rural change in a developed country. Most growth is taking place in areas close to existing urban areas whereas there is decline in more inaccessible areas.

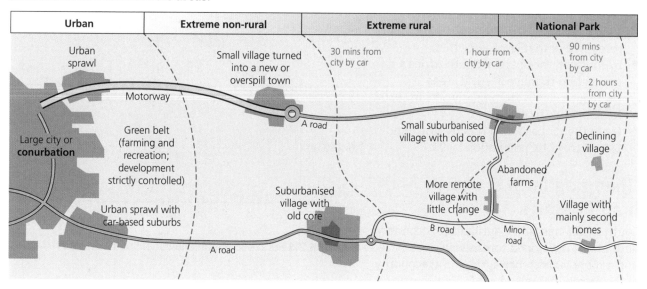

Figure 5.2.2 Rural change and accessibility to large urban areas

The areas showing least change, or decline, are the more isolated areas, i.e. areas that are more than one hour away from large urban areas. This reduces the attractiveness of such areas for commuting. However, there are some areas that are distant from large urban areas that may be attractive for second home owners. These areas include national parks, coastal areas and other tourist areas.

Many rural areas have experienced rural depopulation due to the mechanisation of agriculture. This has led to a widespread loss of employment opportunities in farming. This is especially true in flat, lowland areas where mechanisation is more common. Due to depopulation, many rural services (schools, post offices) have closed, making some rural areas even less attractive for residents. This is a form of negative multiplier effect.

Some rural areas have grown as a result of leisure and tourism. This is largely due to the demand from mobile, affluent populations for recreational activities such as walking, hiking, fishing.

Close to large urban areas, rural settlements have grown due to the demand for more housing and the desire to live in a more attractive environment. Urban areas have grown at their edges (suburbanisation) and some people have moved into smaller settlements (counter-urbanisation).

Conurbation A large-settlement formed by the merging of two or more cities.

Second homes Properties owned by people (generally living in cities) who use them only at the weekend or during holidays.

Negative multiplier effect A downward vicious circle, e.g. a decline in one factor causes a decline in another which in turn reinforces the original decline.

Counter-urbanisation The movement of people from large urban areas to smaller urban areas and rural areas.

Exam tip

Be sure to show that as transport develops, some formerly remote rural areas have become more accessible and have experienced large-scale changes.

Now test yourself

TESTED ⦿

4 Explain the term 'negative multiplier effect'.

5 Explain why some remote rural settlements have grown and some have declined.

6 Explain the term 'extreme non-rural' as shown in Figure 5.2.2.

Revision activity

For your home region/country, identify a large urban area, a rural area surrounding it (with named rural settlements) and a rural area that is still mainly used for farming (named settlements).

5 Rural environments

My Revision Notes: Pearson Edexcel International GCSE Geography

Rural change in an emerging country

Population

+ South Africa's rural population increased steadily from just under 10 million in 1960 to around 20 million in 2000 but has remained at that level since then.
+ In contrast, the share of South Africa's population that live in rural areas decreased slowly from nearly 54% in 1960 to about 50% by 1985.
+ Since then it has fallen rapidly to around 33% of the total population by 2016.

Changing farm economy and land holdings

South Africa has a dual agricultural economy, with a well-developed commercial sector and a more subsistence agricultural economy in the former homeland areas. The contrast between the two sectors could hardly be greater – commercial farms produce 90% of the income from farming but most of the jobs are in black subsistence farming.

+ Commercial farming is dominated by white farmers, although their numbers are decreasing.
+ In the 1950s, there were over 100,000 commercial farmers – now there are about 40,000.
+ Black subsistence farming is concentrated in the former 'homeland' areas.
+ These areas have experienced falling crop yields due to several reasons: overcrowding, use of poor land, overgrazing, soil erosion, emigration of younger workers.

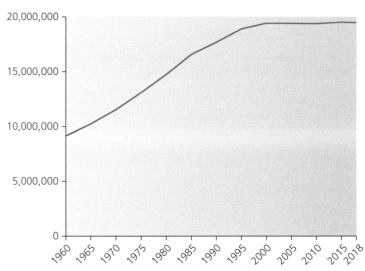

Figure 5.2.3 Rural population change in South Africa, 1960–2018

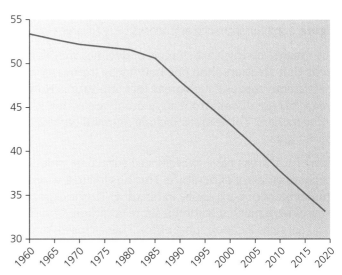

Figure 5.2.4 Rural population as a percentage of the total population

Table 5.2.1 Land per person in South Africa, 1970 and 2020

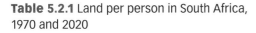

	1970	2020
Cultivated land (ha)	0.6	0.2
Other (ha)	5.5	1.5

> **Homeland** Areas (in South Africa) that were set aside for the black population during the Apartheid era (Apartheid was a form of racial separation forcing different racial groups to live in set areas).

Natural hazards

South Africa experiences a number of natural hazards including flooding, drought and desertification.

+ Drought is a lack of water over a sustained period of time caused by a combination of low rainfall and high evaporation rates.
+ Drought is sometimes called a 'sleeping' hazard as it takes a long time to develop and is not visible at first.
+ Drought is a greater problem in drier areas as there are fewer reserves of water.
+ In 2017 the Western Cape experienced its worst drought for over a century.
+ Desertification occurs when desert features and processes gradually creep into an area.

+ This may be due to natural changes such as reduced rainfall, or it can be due to human activities such as vegetation removal for fuelwood, shelter and fencing.

Rural–urban migration

The migration of mainly young, innovative and skilled workers deprives rural areas of their greatest asset. This prolongs the cycle of poverty in rural areas as those with the skills and knowledge to help the development of the area have left it.

+ Out-migration of the young and skilled means that rural areas have a shortage of skilled workers.
+ Instead, there are larger concentrations of older workers, unskilled workers and poor households.
+ Urban areas are perceived as being more attractive for employment opportunities, so those who can afford to move do so.
+ With younger people moving away, birth rates decline (due to an ageing population) and rural areas fall further behind urban regions.

Now test yourself   TESTED ⬤

7 Outline reasons why population growth in South Africa's rural environments has slowed.
8 Suggest contrasting reasons for the declining productivity of black subsistence farming.
9 Outline the negative impacts of out-migration to South Africa's rural environments.

5.3 Rural environments need to adapt to be socially, economically and environmentally sustainable

Diversification   REVISED ⬤

Diversification refers to the variety of farming and non-farming activities that farmers adopt to make a profit. Diversification is a recent trend but is increasingly important. It developed as the costs of farm inputs increased more than the price received for farm products. Hence the economic profit from farming was being reduced. Diversification allowed farmers to increase

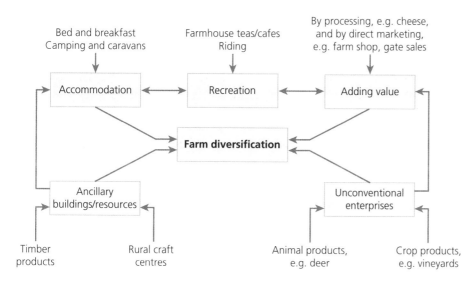

Figure 5.3.1 Options for farm diversification

their earnings from alternative sources. It is widespread in the UK, especially in farms near to urban areas, e.g. Millets Farm near Abingdon, which grows and sells food, has a pick-your-own, fishing, falconry, a farm shop, garden centre, conference and wedding centre.

Diversification requires a number of conditions:
+ availability of capital
+ proper marketing and advertising.

Examples include GM (genetically modified) crops, i.e. those that can produce large increases in yield given optimum conditions; specialist crops and food, e.g. reindeer or ostrich herding), organic farming, i.e. farming without pesticides, herbicides or chemical fertilisers, and recreation and leisure, e.g. farm tours and pick-your-own.

Now test yourself TESTED

1 Outline a) one advantage and b) one disadvantage of farm diversification.
2 How may the diversification options for farmers vary between a farm on the urban fringe and one in a remote rural area?
3 Describe the main characteristics of organic farming.

Achieving rural sustainability in the Eastern Cape, South Africa REVISED

There are several ways in which rural sustainability may be achieved.
+ Soil quality could be improved by maintaining cover crops, using more indigenous species, zero tillage and soil conservation. The key is to reduce the amount of pressure on the soil.
+ Air quality could be improved by ensuring cooking with fuelwood does not occur indoors. It would be better to use more renewable forms of energy (solar and wind, for example) than relying on fuelwood.
+ Water supplies could be protected by having more water butts or, on a larger scale, sand/gravel dams. These would both increase the amount of water captured and reduce evaporation rates.
+ Crop yields could be increased by using organic fertilisers and irrigation. In some cases, indigenous crops such as the khaki bush and aloes have been cultivated to produce essential oils.
+ There are many health problems in the Eastern Cape, leading to low life expectancy, high infant and maternal mortality rates, and widespread malnutrition. Health care needs to be preventative rather than curative and should focus on promoting health, e.g. growth monitoring and provision of immunisation.

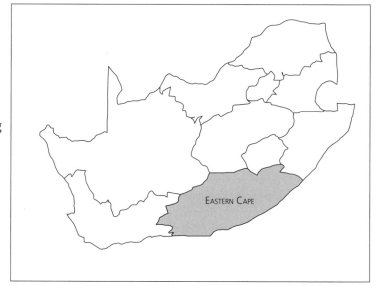

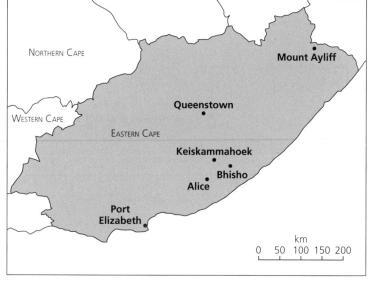

Figure 5.3.2 The location of the Eastern Cape in South Africa

- Poverty is endemic in the area. The Eastern Cape Vision 2030 aims to double agricultural employment by 2030 and increase its share of GDP to 10% from 2.5%. This will require widespread land redistribution to poorer households and greater access to agricultural support.
- Housing is a major problem. Much of the housing is of low quality, made with sub-standard materials and poorly constructed. Pressure needs to be put on national and local governments to increase the provision of affordable, acceptable housing.

Revision activity

Divide the strategies for making rural living more sustainable into those that deal with the physical environment and those that deal with the human environment.

Now test yourself   TESTED ◯

4   Suggest the advantages of preventative health care over curative health care.
5   Outline the likely problems in making rural areas sustainable.
6   Explain how soil quality in the Eastern Cape could be improved.

Exam tip

A case study must have local details – you will not score highly if you just refer to the national scale.

The role of different groups in managing rural challenges

REVISED ◯

The Eastern Cape is home to about 15% (over 6 million) of South Africa's population. Almost 50% of the Province's population is unemployed compared to the national figure of 25%. The majority of people of the Eastern Cape are more rural and significantly poorer than in other parts of South Africa, with a large proportion of the population being reliant to some degree on natural resources for direct subsistence use or indirectly as a form of income generation.

There are many interested parties (stakeholders) in the development of the Eastern Cape Province. These include, among others, the South African government, the Eastern Cape provincial government, international government organisations, such as the European Union, non-government organisations such as Fort Hare University, individual companies, such as Amadlelo Agri, and local communities.

National level – South African government

In 2009, the South African government established the Ministry of Rural Development and Land Reform. Its main objective is to deliver agrarian transformation.

The objectives of the agrarian transformation strategy are widespread and include:
- Social mobilisation to enable rural communities to make rural communities more independent and take initiatives.
- Creating sustainable settlements with access to basic services and economic opportunity; meeting of basic human needs; and the development of infrastructure.
- Establishment of cooperatives and enterprises for economic activities; wealth creation.
- Development and diversification of non-farm activities for strengthening of rural livelihoods.
- Skills development and job creation (youth, women, people living with disabilities).

Agrarian transformation
Change in agricultural practices leading to improved output and increased farm incomes.

Provincial level

The Eastern Cape provincial government manages the region at a local level. Its Vision 2030 has a number of key objectives, including:
- a growing, inclusive and equitable economy
- an educated, empowered and innovative population
- a healthy population

71

+ vibrant, equitably enabled communities, and
+ capable, conscientious and accountable institutions.

University of Fort Hare

The University of Fort Hare, at Alice, has an outreach programme which aims to improve quality of life, and sustainable development. The Eastern Cape has been expanding its dairy production and there are plans to produce 30 million litres of milk annually.

Amadlelo Agri

Amadlelo Agri was established in 2004 by a group of commercial farmers with the goal of transforming the dairy sector. With the Eastern Cape government, it helped develop the first dairy enterprise.

There have been several successful projects including:
+ Fort Hare Dairy Trust, an 800-cow dairy farm and training unit operating jointly with Fort Hare University
+ Seven Stars Trust, a 2000-cow dairy farm, a joint operation with Keiskammahoek landowners
+ Middledrift Dairy Ltd, a 550-cow dairy farm, a joint operation with NEF and the Middledrift community.

Charities

A number of charities operate in the region. For example, the Africa Rural Development Trust (ARDT) is committed to nation building and helping improve the quality of life and development of the weaker sections of society. The ARDT was involved in the construction of a school for children with learning difficulties and physical disabilities at Mount Ayliff in the Eastern Cape region.

The Nolitha Special School caters for children with special needs in three small brick buildings and two wood and iron shelters. These buildings were constructed by the local community. However, the conditions are inadequate. There are limited sewage disposal systems and just a single tap for the supply of fresh water.

Local communities

There are several ways in which local communities have become involved in the management of development issues in the Eastern Cape. The construction of the Nolitha Special School is a good example, as is the restoration of many streams and rivers through the removal of algal blooms and invasive alien species. This has partly been done with the European Union's project 'Protecting South Africa's water for people and biodiversity'. In the Mzimvubu catchment, decades of mismanagement had led to land degradation, overgrazing and many invasive species in streams and rivers. A critical ecosystem partnership fund (CEPF) was established between local communities, NGOs and IGOs (who helped with the funding). The removal of invasive species has led to improved grazing land, better water quality, the establishment of trails and wildlife-viewing hides. Not only have the prospects for farming improved but there is also potential for the development of tourism, and the engagement of some local households in the tourist sector.

Outreach Providing services to people who would otherwise not have access to those services.

Alien species Species that are not native to an area.

NGOs Non-government organisations, ranging from universities, church groups, charities, workers' groups, volunteers.

IGOs International government organisations consisting of many nations (also called multi-government organisations – MGOs).

Revision activity

Make a spider diagram showing all the different groups of people involved in managing a named rural area. Add some detail to show how they are getting involved in managing the area.

Now test yourself

TESTED

7 Explain two ways in which the University of Fort Hare has helped manage rural issues in the Eastern Cape.
8 State the main aims of the Eastern Cape Vision 2030.
9 Briefly explain two ways in which local communities have contributed to the management of rural issues in the Eastern Cape.

Exam tip

It is possible to achieve sustainable development in rural areas but out-migration of young, ambitious people makes it much harder to achieve.

Answers to Now Test Yourself and Exam Practice questions at **www.hoddereducation.co.uk/myrevisionnotesdownload**

Exam practice

1 Arrange the characteristics of a farming system in the correct order. (1 mark)
   A processes, inputs, outputs
   B inputs, outputs, processes
   C inputs, processes, outputs,
   D outputs, inputs, processes

2 Counter-urbanisation refers to: (1 mark)
   A the decline of urban areas
   B the movement of people from large urban areas to rural areas and smaller settlements
   C movement of people from rural to urban areas
   D the growth of urban areas at their edges

3 Approximately what percentage of the Eastern Cape's adult population is unemployed? (1 mark)
   A 10%   C 50%
   B 25%   D 75%

4 Explain why indoor air quality is very poor in many rural households in emerging/developing countries. (2 marks)

5 Define the term 'biome'. (1 mark)

6 Explain the changes in the rural population in a named developing/emerging country. (4 marks)

7 Suggest reasons for the growth of telecommuting. (3 marks)

8 Explain the reasons for the success of a named stakeholder in one project associated with the management of rural issues. (4 marks)

9 Assess the value of ecosystem goods and services provided by tropical rainforests. (8 marks)

Total: 25 marks

Summary

+ There are many types of biomes with different characteristics and distributions.
+ Natural ecosystems provide many goods and services.
+ Humans have changed many ecosystems into farming systems.
+ Rural landscapes vary in terms of landscape, climate, settlement, population, land use, employment, accessibility and management.
+ Many factors lead to change in rural environments, e.g. isolation, declining employment, tourism, suburbanisation, counter-urbanisation and the negative multiplier effect.
+ Rural change in emerging countries includes population growth, economic change, natural hazards and rural–urban migration.
+ Farming has diversified to provide larger incomes for farmers.
+ There are many ways to make rural areas more sustainable – improving soil and water quality, increased crop yield, improved housing and employment.
+ There are many stakeholders in managing challenges in rural environments.

6 Urban environments

6.1 A growing percentage of the world's population lives in urban areas

Contrasting trends in urbanisation

Urbanisation is the process by which an increasing percentage of a country's population comes to live in towns and cities. It may involve rural–urban migration, natural increase and the reclassification of rural settlements as they are engulfed into an expanding city.

> **Urbanisation** An increase in the proportion of people living in urban areas.

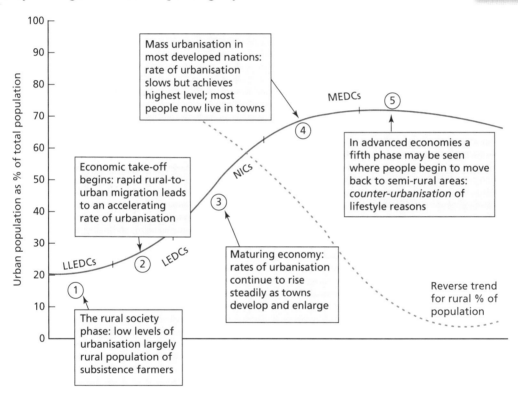

Figure 6.1.1 The process of urbanisation (LLEDCs – Least less economically developed countries; LEDCs – Less economically developed countries; NICs – Newly industrialised countries; MEDC – More economically developed countries)

Suburbanisation is the outward growth of towns and cities to engulf surrounding villages and rural areas. This may result from the out-migration of population from the inner urban areas to the suburbs or from inward rural–urban movement.

Counter-urbanisation is a process involving the movement of population away from inner urban areas to new towns, commuter towns or villages on the edge or just beyond the city limits/rural–urban fringe.

> **LEDCs** Less economically developed countries.
>
> **MEDCs** More economically developed countries.
>
> **Suburbanisation** The growth of urban areas at their edges.
>
> **Counter-urbanisation** The movement of people from large urban areas to smaller urban areas and rural settlements.

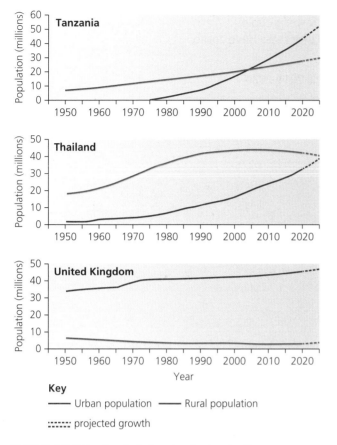

Figure 6.1.2 Contrasts in levels of urbanisation

Revision activity

On a copy of Figure 6.1.1 place the three countries in Figure 6.1.2 at the appropriate point on the graph.

Exam tip

Remember that any model is a simplification – we would not expect all countries to follow this exact pattern.

LLEDC Least less economically developed countries.

Now test yourself TESTED ○

1 Distinguish between urbanisation and counter-urbanisation.
2 Contrast the trends in levels of urbanisation in Figure 6.1.2.
3 Define the term 'least less economically developed countries' (**LLEDCs**).

Rapid urbanisation and the growth of megacities

REVISED ○

There are several reasons for rapid urbanisation in LICs and NICs. These include:

+ The prospects of finding employment, better paid jobs and more secure jobs in urban areas (economic pull factor).
+ Better provision of education and health facilities in urban area (social pull factor).
+ Fewer economic opportunities in rural areas – farming can be low paid, insecure and subject to climate and natural hazards (economic and physical push factors).
+ Poor access to clean water and sanitation, health care and education in rural areas (social push factors).

Consequently, rural–urban migration may be large scale, leading to rural depopulation and the growth of population in urban areas. The majority of those who migrate are young adults.

A megacity is a city with over 10 million people. By 2018, there were 33 cities with a population over 10 million and of these, 27 were in developing/emerging countries.

Megacity growth may be slowing down. Several factors help explain this:

LICs Low-income countries.

NICs Newly industrialising countries.

Megacity A city with over 10 million residents.

75

+ In many cities in the developing world, slow economic growth (or economic decline) has attracted less investment and fewer people.
+ Lower rates of natural increase have occurred, as fertility rates have come down.

Table 6.1.1 Population growth (millions) in selected megacities, 2014–2030

Megacity	Population, 2014 (m)	Population (projected), 2030	Percentage increase/decrease
Tokyo	37.8	37.2	-2%
New Delhi	25	36	+45%
Shanghai	23	30.8	+34%
Mumbai	20.7	27.8	+34%
Beijing	19.5	27.7	+42%
Dhaka	17	27.4	+61%
Karachi	16.1	24.8	+54%
Mexico City	20.8	23.9	+14.5%
São Paulo	20.8	23.4	+13%
New York	18.6	19.9	+7%
Calcutta	14.8	19.1	+29%
Buenos Aires	15	17	+13%
Manila	12.8	16.8	+31%
Rio de Janeiro	12.8	14.2	+11%
Los Angeles	12.3	13.3	+8%

> **Revision activity**
>
> Work out the percentage of megacities in Table 6.1.1 that are in MEDC/LEDCs and emerging nations.

> **Now test yourself**
> TESTED ◯
>
> 4 Outline the reasons for the rapid growth of urbanisation.
> 5 Define the term 'megacity.'
> 6 Identify from Table 6.1.1 the megacity with a) greatest relative increase b) largest absolute increase and c) population decline.

> **Exam tip**
>
> The table only shows selected megacities from 2014. By 2030 there could be new megacities.

Problems associated with rapid urbanisation

REVISED ◯

The rapid growth of urban areas around the world has been one of the most important geographical phenomena of the late twentieth and early twenty-first centuries. For individuals and families, urban areas offer the prospects of a job, a home and an opportunity to improve their standard of living and quality of life. For some, migration to urban areas improves their standard of living, but for others migration may result in unemployment, poor quality housing and deprivation. Rapid urbanisation is associated with problems of congestion, transport, employment, crime and the environment.

Congestion

Rapid urbanisation can lead to large-scale congestion of people (and economic activity). For example, in Mumbai, up to 1 million people live in the Dharavi slum, an area that covers about 2 km². Some 99% of houses do not have a private toilet. Such conditions of congestion fuel the spread of diseases, such as COVID-19 in 2020.

Transport

Congestion is a problem due to vast numbers of cars on the road, and the poor quality/size of roads in many cities. Urban traffic congestion varies with days of the week, time of day, weather and the seasons. Travel is more congested on weekdays, especially during the peak flow times in the morning and evening, i.e. getting to and from work/school.

Congestion may be related to festivals, large sporting events and national holidays. By contrast, during the summer, congestion may decrease as more people walk/cycle to work and schools are closed.

Employment, crime and environmental issues

Most migrants are drawn to large cities by the prospect of employment and a better standard of living. For some this happens, but many are faced with unemployment, underemployment (working only occasionally or just a few hours a day/week), low pay and a lack of job security. Many are forced to enter the informal sector, the unregulated economy with mainly casual jobs, e.g. selling food on the street and domestic service.

In many large urban areas, crime is a problem. This may be partly related to large-scale unemployment and the lack of job opportunities. Often crime is concentrated in areas of high population density. For example, in Islamabad, an informal settlement in Zanjan in Iran, crime is highly concentrated. The main criminal activities include violence, drug trafficking and drug abuse. In Kaduna, northern Nigeria, burglary and stealing are highest in areas of high population density, whereas in wealthier areas car theft and damage are more common.

Environmental issues are widespread. Waste products and waste disposal is a major problem. Some 25% of urban dwellers in LICs have no adequate sanitation and no means of sewage disposal. Air pollution is common in cities in rapidly industrialising countries. Delhi has some of the worst air quality in the world, and poor air has been linked to higher rates of death. Water pollution is widespread as water is used as a dumping ground for agricultural, industrial and domestic waste and for untreated sewage. Water shortages have become more frequent in some cities – overuse of groundwater has led to subsidence in Bangkok. For some people, not being connected to a water tap means that they have to buy water from sellers, and that is very expensive.

> **Revision activity**
>
> Make two lists: one with the advantages of moving to an urban area and one for the disadvantages of moving to urban areas.

> **Exam tip**
>
> Make sure you provide named support when answering questions about problems associated with rapid urbanisation.

Now test yourself   TESTED ◯

7   Estimate the population density of Dharavi.

8   Suggest why there are variations in traffic congestion.

9   Explain why poor people may pay more for their water than rich people.

6.2 Cities face a range of social and environmental challenges resulting from rapid growth and resource demands

Urban land use patterns   REVISED ◯

Urban land use refers to activities such as industry, housing and commerce that may be found in towns and cities. Land values decrease with distance from the city centre. Land values also increase with accessibility to good transport routes.

> **Land use** The main function for which an area is used, e.g. residential, industrial, commercial.
>
> **Accessibility** How easy it is to get to a place.

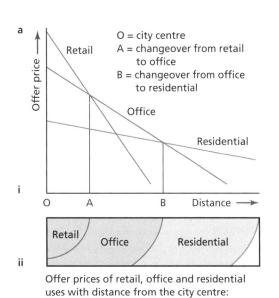

a

i

O = city centre
A = changeover from retail to office
B = changeover from office to residential

ii Offer prices of retail, office and residential uses with distance from the city centre:
i section across the urban value surface
ii plan of the urban value surface

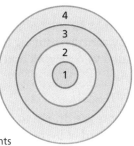

b Concentric zone model (Burgess, 1925)

- model based on Chicago in the 1920s
- the city is growing spatially due to immigration and natural increase
- the area around the CBD has the lowest status and highest density housing
- residents move outwards with increasing social class and their homes are taken by new migrants

Key

1 CBD (central business district)
2 Zone in transition/light manufacturing
3 Low-class residential
4 Medium-and high-class residential

Figure 6.2.1 Bid rent theory and the concentric zone model

The central business district (CBD) is where most of the commercial activity is found. It is the most accessible (to public transport) and has the highest land values. It tends to have high rise buildings owing to the strong demand for land, but a shortage of space.

> **Central business district (CBD)** The main commercial part of a city, generally found in the most central area.

Core and frame elements of the CBD

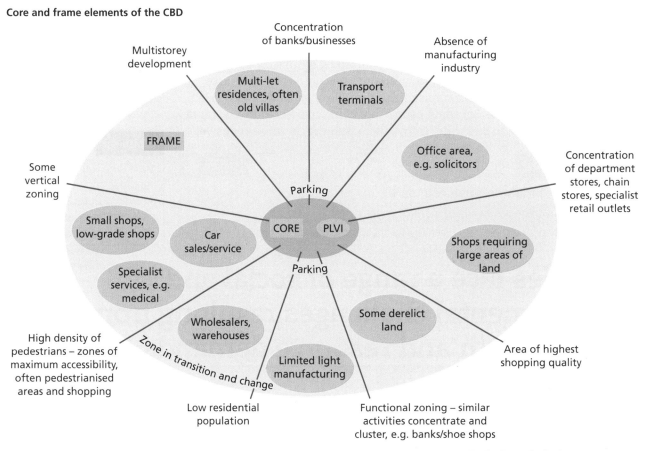

PLVI = peak land value intersection: the highest rated, busiest, most accessible part of a CBD

Figure 6.2.2 Core and frame characteristics of the CBD

Most **residential** areas are found in the suburbs. The suburbs refer to the outer part of an urban area. Suburbs generally consist of residential housing and shops of a low order (newsagents, small supermarkets).

> **Suburbs** The outer parts of a city, usually with a residential function.

In contrast, the rural–urban fringe is the boundary of a town or city, where new buildings are changing land use from rural to urban.

Industrial areas occur in several locations such as the inner city (the area surrounding the CBD), along major transport routes, and in edge-of-town locations. In many cities, the inner city is the older industrial area of the city and may suffer from decay and neglect, leading to social problems. Inner cities are characterised by poor quality terraced housing with old manufacturing industry nearby.

However, urban areas are changing rapidly. Much retailing and commerce is now taking place on the edge of towns and inner city areas are being used for residential purposes.

> **Rural–urban fringe** The edge of a city/boundary between urban and rural functions.

> **Revision activity**
>
> For a named urban area in your home region, identify an industrial area, a suburb and a development on the rural–urban fringe.

> **Now test yourself** TESTED ◯
>
> 1 Define the term 'accessibility'.
> 2 Suggest contrasting reasons for the location of industrial areas in cities.
> 3 Explain how and why land values vary in a city.

> **Exam tip**
>
> No city shows exactly the same pattern as the concentric zone model, but there are many similarities, especially on large urban areas in developed countries.

Urban challenges: developed countries REVISED ◯

All urban areas face many challenges, even prominent cities, such as London, UK.

+ One of these is to provide food for their residents. There are complex global supply chains, which can be badly disrupted. For example, London's food sector accounts for £20 billion and about 10% of jobs in the city. Nevertheless, there are many food banks in London, helping to feed an estimated 1.5 million people who go hungry.
+ Most of London's energy comes from gas. Although there is a plan to produce 15% of London's energy from renewable sources by 2030, at present less than 1% comes from renewables.
+ Transport faces issues regarding congestion and sustainability. Central London has a 20 mph speed limit and a congestion charge to limit the number of car trips made. However, the number of car journeys is increasing. This reduces the reliability of bus journeys due to increased congestion. London has a well-developed underground network, although some parts of it have very poor air quality.
+ London produces a huge amount of waste. Some of this is disposed of in landfill sites, some is burnt to generate electricity and some is recycled. On average most households produce around 1000 kg of waste yearly.
+ London has concentrated resource consumption due its size and wealth. Its ecological footprint is 6.6 global hectares (gha) per person, compared with a world average of 2.8 gha and the UK's overall figure of 6.3 gha.

London is a multi-cultural city. However, based on 2011 census data, segregation is evident. There is a higher concentration of white British towards the edge of the city. In contrast, some minority ethnic groups are concentrated nearer the centre, such as Bangladeshis (in the east end). Indians and Pakistanis are more likely to be located in the west, the east end and some parts of south London.

> **Food bank** A charity providing food for those who cannot afford to buy sufficient food to meet their needs.

> **Landfill** Dumping of waste materials in the ground.

> **Ecological footprint** A measurement of the amount of land needed to provide a population with the resources it consumes (water, food, energy). It is measured in global hectares (gha).

> **Multi-cultural** An area with many different races and socio-economic groups.

> **Segregation** Keeping apart people of different backgrounds.

> **Revision activity**
>
> For a named urban area in your home region, identify two or more urban challenges and write a sentence on each, stating what the challenge is.

> **Now test yourself** TESTED ◯
>
> 4 Suggest why air quality in London varies spatially and temporally.
> 5 Define the term 'ecological footprint'.
> 6 Suggest why London has a high ecological footprint.

Urban challenges in a developing country

The total number of slum dwellers in the world stood at about 980 million people in 2015. This represents about 32% of the world's urban population, but 78.2% of the urban population is in LICs.

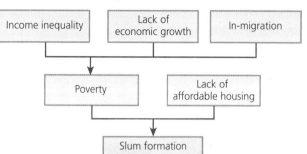

Exam tip

Urban challenges vary from city to city, and in different parts of a country.

Figure 6.2.3 Inequality, poverty and slum formation

Mumbai experiences many of the problems resulting from rapid city growth – poverty, unemployment and underemployment, limited access to health care and education, poor sanitation and access to electricity.

+ Most of the residents live in slums/squatter settlements, and these have limited security of tenure.
+ Dharavi, the main slum in South Mumbai, is an area of about 2 km², and home for up to 1 million people.
+ Due to its close proximity to Mumbai's financial and commercial district, there is great pressure to clear parts of Dharavi for modern developments.

Squatter settlements
An illegal settlement where homes have been built by the residents.

Table 6.2.1 The positives and negatives of living in a slum

Positive aspects	Negative aspects
+ They are points of assimilation for immigrants. + Informal entrepreneurs can work here and have clients extending to the rest of the city. + Informal employment, based at home, avoids commuting . + There is a strong sense of kinship and family support. + Crime rates are relatively low.	+ Security of tenure is often lacking. + Basic services are absent, especially water and sanitation. + Overcrowding is common. + Sites are often hazardous. + Levels of hygiene and sanitation are poor, and disease is common.

Informal economy

+ The formal economy refers to the regulated economy, e.g. offices, factories, shops and services such as health care, education and government.
+ Much of the formal economy produces goods and services for wealthy people.
+ By contrast, the informal economy is small scale, locally owned and labour intensive.
+ Dharavi has many informal activities that provide a livelihood for many of its residents.
+ Up to 85% of Dharavi adults work locally, and there are major recycling industries and pottery industries.
+ Working conditions for the recycling industry can be very dangerous.

Informal economy The unregulated, untaxed economy – sometimes called the 'black market'.

Water and sanitation

+ The Dharavi slum has a poor sewage and drainage system.
+ The area is subject to floods in the wet season.
+ There are up to 4000 cases per day of diphtheria and typhoid, partly the result of a lack of a proper sewer system.
+ Water access is from standpipes. Access to water is limited and many pumps are only available for two hours per day.

- Many people used the Mahim creek for washing but it is also used for urination and defecation.
- Open sewers drain into the creek, bringing a range of pollutants. Dharavi has a very limited number of toilets – about one for every five hundred people.

Air quality

- Air quality in Dharavi (and Mumbai in general) is poor due to industrial and vehicle emissions, open burning and dust.
- Air quality is generally worse in winter (November to February) as there is less rainfall to clear the air.

Quality of life

- Many aspects of Dharavi lead to a low quality of life, e.g. lack of security of housing (up to 90% may be illegal), low life expectancy (about 50 years), low wages (on average between £1 and £5/day), poor air quality and the risk of disease.
- Nevertheless, there are many jobs available and Dharavi's informal economy is worth between US $500 million and $1 billion a year (£250 million–£500 million).

> **Revision activity**
>
> Make a spider diagram showing the challenges that are found in an urban area in a developing/emerging country.

> **Now test yourself**   TESTED ◯
>
> 7   Outline one advantage and one disadvantage of Dharavi's location.
> 8   Define the term 'informal economy'.
> 9   Explain why water pollution is a problem in Dharavi.

> **Exam tip**
>
> Make sure that your case studies have specific geographic facts and features.

6.3 Different strategies can be used to manage social, economic and environmental challenges in a sustainable manner

Development of the rural–urban fringe   REVISED ◯

The rural–urban fringe is the area at the edge of a city where it meets the countryside. There are many pressures on the rural–urban fringe. These include:

- more housing, e.g. Blackbird Leys, Oxford
- industrial growth, e.g. Oxford Science Park
- transport infrastructure, e.g. M25, London
- recreational pressures for golf courses and sports stadia, e.g. Kassam Stadium, Oxford.

> **Rural–urban fringe** The edge of a city/boundary between urban and rural functions.

Table 6.3.1 The advantages and disadvantages of out-of-town (rural–urban fringe) shopping centres

Advantages	Disadvantages
Plenty of free parking	They destroy large amounts of undeveloped, valuable habitats
Lots of space so shops are not cramped	They lead to pollution and environmental problems at the edge of town
Easily accessible by car	They only help those with cars
Developments on the edge of town reduce the environmental pressures and problems in city centres	Successful out-of-town developments may take trade away from city centres and lead to a decline in sales in the CBD

81

Greenfield and brownfield sites

A greenfield site is a site that has not previously been developed. Most greenfield sites are on the edge of town, although not necessarily so. A brownfield site is a site that has previously been used and has become derelict.

Table 6.3.2 The advantages and disadvantages of greenfield and brownfield sites

| Greenfield sites | | Brownfield sites | |
Advantages	Disadvantages	Advantages	Disadvantages
+ Land may be accessible	+ Habitat destruction	+ Redevelopment of disused land	+ Land may be contaminated
+ Cheaper land	+ Reduction in biodiversity	+ Does not harm the environment	+ Widespread air and water pollution
+ People prefer more space and pleasant environments	+ Increased impermeability leads to flooding	+ Creates jobs locally	+ Congestion

Exam tip

Brownfield sites can occur anywhere in a city – they occur wherever there is derelict land.

Revision activity

Draw two spider diagrams to show the advantages and disadvantages of out-of-town (rural–urban fringe) developments.

Now test yourself   TESTED ⦾

1 Identify the part of a city most likely to have a greenfield site.
2 State the advantages of developing brownfield sites.
3 Describe the disadvantages of developing greenfield sites.

Sustainable urban systems   REVISED ⦿

Large cities are often considered unsustainable because they consume huge amounts of resources and they produce vast amounts of waste. Sustainable urban development meets the needs of the present generation without compromising the needs of future generations. The Rogers' model (*Cities for a Small Planet*) compares a sustainable city with that of an unsustainable one. In the sustainable city, inputs are smaller and there is more recycling.

Linear cities consume a large volume of resources and pollute on a large scale

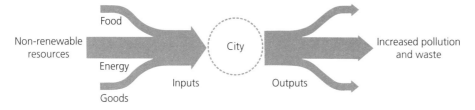

Circular cities consume resources on a smaller scale and cause less pollution

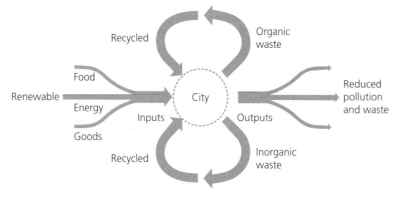

Figure 6.3.1 Linear (non-sustainable) and circular (sustainable) cities

To achieve sustainability, a number of options are available:

+ reduce the use of fossil fuel, e.g. by promoting public transport
+ keep waste production to within levels that can be treated locally
+ provide sufficient green spaces
+ re-use and reclaim land, e.g. brownfield sites
+ encourage active involvement of the local community
+ conserve non-renewable resources
+ use renewable resources.

Compact cities minimise the amount of distance travelled, use less space, require less infrastructure (pipes, cables, roads etc.), are easier to provide a public transport network for, and reduce urban sprawl.

But if the compact city covers too large an area, it becomes congested, over-crowded, over-priced and polluted. It then becomes unsustainable.

Figure 6.3.2 Developing sustainable urban areas

A sustainable future requires:

+ use of appropriate technology, materials and design
+ acceptable minimum standards of living
+ social acceptability of projects
+ widespread public participation.

The main dimensions of sustainable development are:

+ provision of adequate shelter for all
+ improvement of human settlement management
+ sustainable land use planning and management
+ integrated provision of environmental infrastructure: water, sanitation, drainage and solid waste management
+ sustainable energy and transport systems
+ settlement planning in disaster-prone areas
+ sustainable construction industry activities
+ meet the urban health challenge.

> **Now test yourself**   TESTED ◯
>
> 4 Describe the main characteristics of the linear city.
> 5 Outline how the circular city is sustainable.
> 6 Outline two social demands in the sustainable city.

Revision activity

Using Figure 6.3.2, choose three options which you feel are the most important for achieving sustainable urban development. Justify your choices.

Exam tip

Use examples of sustainable urban development from your home region and comment on how successful they have been.

Linear city A city with large inputs of resources and large amounts of waste

Circular city A city with a high proportion of renewable resources and recycling

---

Stakeholders in managing urban areas   REVISED ◯

A stakeholder refers to any group or individual who is affected, or can affect, or has an interest in a development, e.g. new housing or a new out-of-town development. Some stakeholders may be supportive of the development, some may be against it, and others may see both advantages

83

and disadvantages of the development. Stakeholders include the individuals involved in a scheme, national and local governments, charities, local churches, local planners, building companies, property owners, retailers, developers and estate agents.

For example, the Houldsworth Village Partnership in Manchester involves at least 60 different stakeholder groups including the Stockport Metropolitan Council, the Guinness Northern Counties Housing Trust, the University of Manchester, St Elisabeth's School, existing residents and traders, Stockport Sports Trust and the Reddish Crime Panel.

Some stakeholders may have conflicting interests, e.g. established residents in areas desiring no new developments compared with younger people desiring new, affordable housing developments. Corporate objectives may differ from individual ones.

The impact of a new development may have negative impacts on others, e.g. an out-of-town development may create jobs locally but also lead to an increase in congestion, air pollution and a decline in habitat.

Stakeholders may have to compromise their individual beliefs and values for the greater good. However, it is possible for stakeholders with an interest in the success of a partnership/development to attempt to minimise the impact on stakeholders with negative views, e.g. a new housing development retaining open space and vegetation rather than having high density housing only.

Now test yourself

TESTED

7 Explain the likely views of two contrasting stakeholders (e.g. existing residents and first-time buyers) about a new local housing development.

8 Identify one private and one public stakeholder in the Houldsworth Village Partnership.

9 State one reason why it may be difficult to reach agreement on how best to develop urban areas sustainably.

Exam practice

1 Urbanisation is defined as: (1 mark)
   A the growth of urban areas
   B an increase in the proportion of people living in urban areas
   C an increase in the population living in urban areas
   D all of the above

2 A greenfield site is: (1 mark)
   A an area of land that has never been developed
   B an area of conservation
   C a derelict site
   D an area with edge-of-town developments

3 The proportion of people who live in slums around the world is approximately: (1 mark)
   A 10%
   B 30%
   C 50%
   D 70%

4 Compare push and pull factors in relation to urbanisation. (2 marks)

5 Outline the advantages of compact cities. (3 marks)

6 Describe the main locations in which industry is located in an urban area. (4 marks)

7 Define the term 'squatter settlement'. (1 mark)

8 Suggest reasons for the growth of counter-urbanisation in developed countries. (4 marks)

9 Analyse how urban areas may be made more sustainable. (8 marks)

Total: 25 marks

Answers to Now Test Yourself and Exam Practice questions at **www.hoddereducation.co.uk/myrevisionnotesdownloa**

Summary

✚ There are contrasting global trends in urbanisation, suburbanisation, and counter-urbanisation.

✚ Many factors affect the rate of urbanisation and the emergence of megacities.

✚ Rapid urbanisation is associated with many problems, e.g. congestion, transport, employment, crime and environmental issues.

✚ Factors affecting urban land-use include locational needs, accessibility and land value.

✚ Urban challenges in wealthy countries include food, energy, transport, waste disposal, concentrated resource consumption and segregation.

✚ Urban challenges in emerging countries include squatter settlements, the informal economy, urban pollution and low quality of life.

✚ Development of the rural–urban fringe includes housing estates, retail, business and science parks, and industrial estates.

✚ There are many strategies to make urban living more sustainable, e.g. in waste disposal, transport, education, health, employment and housing.

✚ There are many stakeholders in managing the challenges in urban environments.

7 Fragile environments and climate change

7.1 Fragile environments are under threat from desertification, deforestation and global climate change

Distribution and characteristics of fragile environments

A fragile environment is one that is vulnerable to change and may find it difficult to recover from natural or human-induced changes. Some ecosystems can cope with wide variations in climatic conditions and human pressures, whereas others are much more sensitive to change. Fragile environments include arid and semi-arid environments, rainforests and cold environments (Figures 7.1.1a, b and c).

Some natural events, such as volcanic eruptions, tsunamis, tropical cyclones and extreme weather may cause change, but increasingly it is anthropogenic events (human-induced events) such as deforestation, intensive agriculture and urbanisations that are increasing the pressure on natural ecosystems.

> **Fragile environment** A fragile environment is one that is vulnerable to change and may find it difficult to recover from natural or human-induced changes.
>
> **Deforestation** The removal of forest cover – it can be complete or partial.

Table 7.1.1 Characteristics of fragile environments

Fragile ecosystem	Reasons for fragility
Tropical rainforest	Infertile soils; rapid deforestation; vulnerability to climate change
Coral reefs	Vulnerable to warming oceans, pollution, tourism
Arid and semi-arid environments	Lack of moisture so regeneration is slowed down
Tundra and alpine environments	Low temperatures so regeneration is limited; vulnerability to climate change as habitats are changed

Figure 7.1.1 Fragile environments: a) Semi-arid environment near Uluru, Australia;

Revision activity

Draw a spider diagram to show the threats to coral reefs.

Exam tip

Make sure you can name an example of each type of fragile environment.

b) Submerged rainforest, Batang Ai, Malaysia;

Now test yourself

1 Define the term 'fragile environment'.
2 Suggest reasons why the environments in Figures 7.1.1 a–c may be considered fragile.
3 State **two** natural threats and two human threats to fragile environments.

c) Enjoying the sunshine, Seefeld, Austria

Answers to Now Test Yourself and Exam Practice questions at **www.hoddereducation.co.uk/myrevisionnotesdownload**

Desertification and deforestation

Desertification

Desertification is the spread of desert-like conditions into areas that were previously green. These areas are becoming biologically less productive than they once were and can no longer support as many people at the same standard of living.

There are many inter-related causes of desertification including drought, population pressure, fuel supply (deforestation for fuelwood), overgrazing and migration. Desertification is especially prevalent in sub-Saharan Africa and Central Asia. Up to two billion people – over one-third of the world's population – are at risk of its effects.

> **Desertification** The spread of desert-like conditions into previously productive regions.
>
> **Sedentarisation** Making nomadic farmers remain in one area and not move around.

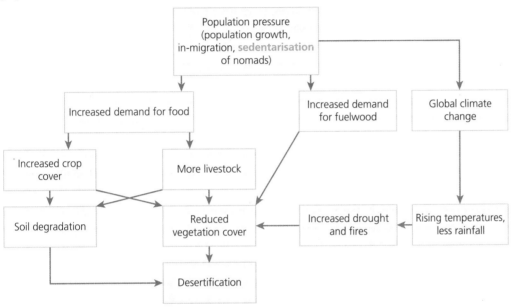

Figure 7.1.2 A model of the causes of desertification

Deforestation

Since agriculture began, almost one-third of the world's forests have been cut down, especially in temperate areas and, increasingly, in areas of tropical rainforest.

Much of the world's forest has been cut down to make way for farming, but some forests have been destroyed due to acid rain. Other forests are cut down for timber, settlement, transport developments and mining, and others are flooded to make way for hydroelectric schemes.

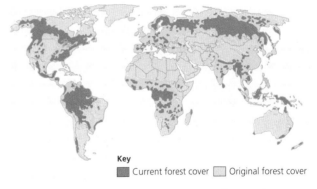

Key
■ Current forest cover □ Original forest cover

Figure 7.1.3 The global distribution of deforestation

Exam tip

Remember that there are many types of deforestation. Clear-felling is the removal of whole trees but pollarding and coppicing remove only part of the tree. There is also selective deforestation where a particular species of tree or a small percentage of the trees are removed.

Revision activity

Draw a simplified diagram to show the process of desertification.

Now test yourself TESTED

4 Define the term 'desertification'.

5 Outline the potential causes of desertification.

6 Describe the global variations in the extent of deforestation.

Natural climate change and the enhanced greenhouse effect

Natural climate change

There are many reasons for long-term changes in the Earth's climate (and they all pre-date the present global climate change), including tectonic movement, mountain building, volcanic activity, solar output, atmospheric dust and changes in the Earth's position relative to the Sun.

The Milankovitch cycles show that the amount of solar energy reaching the Earth varies with changes in the Earth's orbit, its tilt and its 'wobble' – the direction of its rotation. The Earth's orbit varies over a timescale of about 100,000 years. When it is further from the Sun, it receives less energy. In addition, when the tilt is greater, seasons are longer. The 'wobble' determines which hemisphere is facing the Sun – northern or southern.

Other natural causes include volcanic eruptions. Those more likely to cause changes to climate are large eruptions, especially in tropical areas. For example, the eruption of Mt Pinatubo in the Philippines in 1991 led to a drop in mean global temperature of 0.3°C. Even by 2005, the drop in global temperatures due to Mt Pinatubo was 0.1°C. Sunspot activity (solar flaring) occurs on an 11- and 22-year cycle. Atmospheric dust is also believed to block or reflect incoming solar energy, thereby leading to lower temperatures on Earth.

Human causes of climate change

The greenhouse effect is the process by which certain gases (greenhouse gases) allow short-wave radiation to pass through the atmosphere but trap a proportion of outgoing long-wave radiation from the Earth. This leads to a warming of the atmosphere. It is a natural process and vital for life on Earth.

Global climate change
The recent changes in global temperatures due to human activities, also known as global warming/climate crisis/enhanced greenhouse effect.

Milankovitch cycles
Variations in the amount of solar radiation received by the Earth due to variations in the Earth's orbit, its tilt and its wobble.

Greenhouse effect
The process in which certain greenhouse gases allow short-wave radiation to pass through the atmosphere but trap a portion of the outgoing long-wave radiation, thereby raising temperatures.

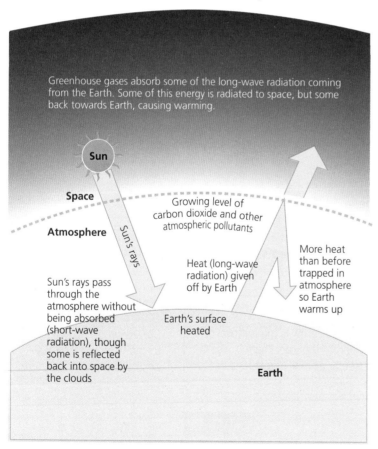

Greenhouse gases absorb some of the long-wave radiation coming from the Earth. Some of this energy is radiated to space, but some back towards Earth, causing warming.

Sun

Space

Atmosphere

Sun's rays

Growing level of carbon dioxide and other atmospheric pollutants

Sun's rays pass through the atmosphere without being absorbed (short-wave radiation), though some is reflected back into space by the clouds

Heat (long-wave radiation) given off by Earth

Earth's surface heated

More heat than before trapped in atmosphere so Earth warms up

Earth

Figure 7.1.4 The greenhouse effect

The most common greenhouse gas is water vapour, which accounts for about 50% of the **natural greenhouse effect**. However, the gases which account for **human causes** of climate change are carbon dioxide (CO_2), methane (CH_4) and chlorofluorocarbons (CFCs).

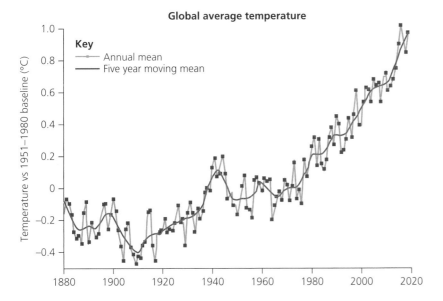

Figure 7.1.5 Changes in the global average temperature relative to the 1951–1980 average

The enhanced greenhouse effect (EGHE)

Atmospheric levels of CO_2 have risen from around 315 parts per million (ppm) in 1950 to over 420 ppm by 2020 and are predicted to rise to 600 ppm by 2050. The rise is due to human activities such as burning fossil fuels (coal, oil and natural gas) and land-use changes such as deforestation. This is known by many terms, e.g. the enhanced greenhouse effect/global climate change/climate crisis/global warming.

7.2 Impacts of desertification, deforestation and climate change on fragile environments

The impacts of desertification   REVISED ○

Up to 12 million hectares of land and 20 million tonnes of grain are lost to desertification every year. One-third of the Earth's land is threatened with desertification. Soil exhaustion decreases world food production, and 20 million tonnes of cereal are lost every year due to desertification.

Table 7.2.1 Some impacts of desertification

Environmental	Economic	Social
+ Loss of soil nutrients through wind and water erosion. + Loss of **biodiversity** as vegetation is removed. + Reduction in land available for crops and grazing land. + Increased sedimentation of rivers and reservoirs due to soil erosion.	+ Reduced income from **pastoralism** and the cultivation of crops. + Decreased availability of **fuelwood**. + Increased rural poverty. + Increased dependence on food aid.	+ Loss of traditional knowledge and skills. + Forced migration due to food scarcity. + Increased rural poverty. + Social tensions between migrants and local people.

Biodiversity The variety of living organisms in an environment including habitat diversity, species diversity and population (genetic) diversity.

Pastoralism A farming system based on rearing of animals.

Fuelwood The use of trees and other vegetation as a source of fuel.

Now test yourself

TESTED ◯

1 Briefly explain why desertification leads to increased soil erosion.

2 Suggest two consequences of the removal of vegetation for fuelwood.

3 Identify the areas most at risk from desertification.

Revision activity

Draw a spider diagram to show the environmental, economic and social impacts of desertification.

Exam tip

Desertification does not occur in deserts – they are already non-productive.

The impacts of deforestation

REVISED ◯

Deforestation has many impacts, including the loss of biodiversity, its contribution to global and local climate change, its impact on economic development and increased rates of soil erosion.

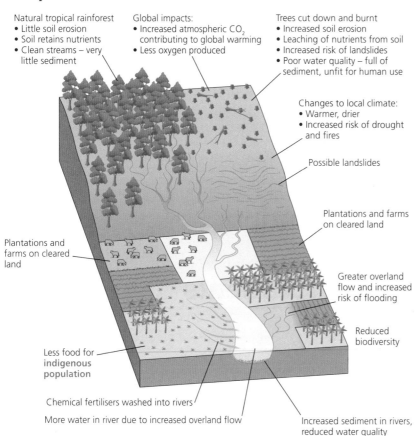

Natural tropical rainforest
• Little soil erosion
• Soil retains nutrients
• Clean streams – very little sediment

Global impacts:
• Increased atmospheric CO_2 contributing to global warming
• Less oxygen produced

Trees cut down and burnt
• Increased soil erosion
• Leaching of nutrients from soil
• Increased risk of landslides
• Poor water quality – full of sediment, unfit for human use

Changes to local climate:
• Warmer, drier
• Increased risk of drought and fires

Possible landslides

Plantations and farms on cleared land

Plantations and farms on cleared land

Greater overland flow and increased risk of flooding

Reduced biodiversity

Less food for indigenous population

Chemical fertilisers washed into rivers

More water in river due to increased overland flow

Increased sediment in rivers, reduced water quality

Figure 7.2.1 Some impacts of deforestation

Indigenous populations Ethnic groups who are the original/earliest known inhabitants of a region, also known as First People/ First Nation, or Aboriginal people, for example.

Answers to Now Test Yourself and Exam Practice questions at **www.hoddereducation.co.uk/myrevisionnotesdownload**

It is thought that the area of tropical rainforests worldwide has halved since 1800 to 1.4 billion hectares. As a result of deforestation, up to 100,000 species may become extinct every year.

Tropical forests are a vital source of material for medicines. Of the 3000 plants that may help cure cancer, 70% are found in tropical forests.

Revision activity

Make a table showing the environmental, economic and social impacts of deforestation.

Now test yourself

TESTED ◯

4 Describe the impacts of deforestation on soils, as shown in Figure 7.2.1.
5 Identify the impacts of deforestation on the water cycle.
6 Briefly explain why deforestation may lead to increased risk of drought and fire.

Exam tip

Some impacts are intentional, e.g. clearing of forested land to make way for farmland. Others are unintentional, e.g. clearing of forests leading to drought and fire.

Impacts of climate change on fragile environments and people

REVISED ◯

Rising sea levels

Sea levels are rising as the oceans warm (the 'steric effect') and as ice on land melts. Even if there was no ice melt, the oceans would increase in size as warmer water expands.

Steric effect The expansion of seawater as it gets warmer, leading to a rise in sea level.

Tens of millions of people are at risk from rising sea levels. Many low-lying islands and many of the world's megacities are less than 10m above sea level and coastal flooding is a major risk. In addition, food production in coastal areas is likely to be affected by saltwater intrusion.

Increase in hazards

Climate change is likely to lead to an increase in hazards. Increased atmospheric energy is predicted to lead to an increase in tropical storms (hurricanes), as well as drought and fire.

In the Amazon rainforest there has been an increase in drought events, such as those of 2005, 2010 and 2015. These have led, in part, to increased tree mortality and a reduction of rainfall over the rainforest. Similarly, in the Middle East and North Africa, water stress is likely to increase due to falling rainfall levels and increased demand for water.

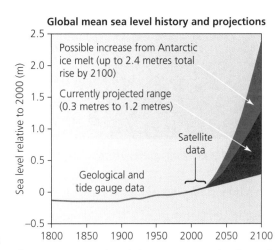

Figure 7.2.2 Sea level rise 1800–2100 (projected)

Ecosystem changes

The effect of climate change is likely to cause latitudinal shifts relative to the Equator and altitudinal shifts as biomes move up-slope. At high altitude and high latitude, biomes may have nowhere to retreat and may become extinct.

There are also changes taking place in the world's oceans. For example, increased ocean temperatures and ocean acidification have led to coral bleaching and a decline in species that build their own skeletons. Fish species are migrating pole-wards into cooler water.

Ocean acidification The decreasing pH due to increasing levels of carbon dioxide absorbed by seawater.

Decreased employment opportunities

Climate change may lead to major job losses in different environments. For example, agriculture may become less productive in areas which become drier and hotter, and tourism in many mountainous areas will decline as glaciers retreat. Fishermen may have to change the species that they catch.

Changing settlement patterns

Settlements in low-lying areas may be abandoned or relocated. Many settlements – or parts of settlements such as those along the USA's east coast – will require considerable protection from rising sea levels and coastal flooding or will have to be abandoned.

Figure 7.2.3 Kivalina, Alaska (USA) – a settlement at risk from climate change

Health and well-being

A rise in temperature of 2°C could expose up to 60 million more Africans to malaria. Mosquitos would be able to breed in areas previously too cool for them. Rising temperatures could lead to increased risk of dehydration, malnutrition and heat stress.

Agricultural crop yields, limit to cultivation and soil erosion

The decline in water resources may make it difficult for some farmers to continue the type of farming they currently practise. They may have to change crop/type of farming and could be forced out of farming altogether.

A rise of 2°C could lead to 200 million more people experiencing hunger, while a rise of 3°C could lead to up to 500 million more people experiencing hunger.

> **Revision activity**
>
> Make a table on the impact of climate change on fragile environments and people. Make three columns – social, economic and environmental – and in each column place the impacts shown in this section.

> **Exam tip**
>
> Remember that there are different scenarios for future climate change – it depends on how much greenhouse gases are released in future and whether there are schemes to tackle global climate change.

Now test yourself   TESTED ◯

7 Outline **two** reasons why sea levels are rising.

8 State the projected range of sea level rises by 2100:
   a) At current levels.
   b) Due to a possible ice melt from Antarctica.

9 Suggest why Kivalina settlement may have to relocate.

Answers to Now Test Yourself and Exam Practice questions at **www.hoddereducation.co.uk/myrevisionnotesdownload**

7.3 Responses to desertification, deforestation and climate change vary depending on a country's level of development

The use of technology to solve water shortages in fragile environments

REVISED ⬤

Desalination

Desalination refers to the removal of salts from seawater to produce fresh water for human consumption and irrigation use. In many dry areas, seawater is a vital potential source of fresh water and can be used to combat desertification.

Due to high energy input, the financial costs of desalinating seawater are generally still high. But alternative water sources are not always available. There are around 20,000 desalination plants operating worldwide, producing almost 90 million m³ of water per day for 300 million people.

Other sources of water in dry areas include groundwater, which could be brought to the surface by using pumps. However, in dry areas groundwater is a finite, non-renewable resource.

Desalination The removal of salts from seawater to produce fresh water.

Groundwater Fresh water stored in underground rocks (aquifers).

Overgrazing The impact of too many livestock feeding on vegetation, leading to a decline in quality and amount of vegetation cover.

Over-cultivation Attempts to grow too many crops over a period of time, leading to a decline in soil quality.

High yielding varieties (HVVs) Genetically modified crops that produce greater yields per hectare, if the soil conditions are correct.

Agroforestry Combining agriculture and forestry.

Table 7.3.1 Other measures to tackle desertification

Cause of desertification	Strategies for prevention
Overgrazing	Improve stock quality; breeds that are adapted to dry conditions; reduction of herd size; use of wider area to reduce grazing pressure.
Over-cultivation	Use of fertilisers can increase yields and reduce the amount of land needed; high-yielding varieties (HYVs) of crops and drought-resistant crops could be introduced; crop rotation, irrigation and zero tillage (not ploughing) can improve soil quality and reduce pressure on soils.
Deforestation	Agroforestry (combining agriculture with forestry), e.g. using trees for fodder, fuel and building. Trees protect, shade and fertilise the soil. Involve village members in managing forests. Use of alternative fuels instead of wood.

Revision activity

Make a table with two columns stating the advantages and disadvantages of desalination.

Now test yourself

TESTED ⬤

1 Explain how desalination may help in the management of desertification.
2 Outline the advantages of using trees to combat desertification.
3 Explain the term 'zero tillage'.

Exam tip

Make sure that you can evaluate attempts to tackle desertification, i.e. state the advantages and disadvantages of each scheme.

Approaches to the sustainable use and management of a named rainforest: the Amazon rainforest, Brazil

Central Amazon Conservation Complex

The Central Amazon Conservation Complex is one of the world's largest conservation areas, covering 53,230 km². It protects several areas with rare, endemic and/or endangered species.

It is the largest protected area in the Amazon and was formed by the merging of the Jau National Park with the Anavilhanas National Park, Amana Sustainable Development Reserve and the Mamairaua Sustainable Development Reserve.

Brazil's Forest Code

The Forest Code is a law that requires landowners in the Amazon to maintain a proportion of their land (80%) as forest. It was passed in 1965 and revised in 2012.

Other initiatives

At a local scale, some indigenous groups use many ways to use the rainforest sustainably. In the Yanesha Forestry Cooperative Project, farmers cut a strip of rainforest some 20–40 metres wide, farm it and then let it recover. The narrow belt allows rapid recovery and secondary forests grow back within twenty years. Other communities enrich their soils by adding animal bones and charcoal. This increases soil fertility and allows them to farm the land more productively. Others plant fig trees on degraded land. These attract birds and bats which bring in seeds from neighbouring forests. The birds and bats deposit the seeds in their droppings thereby helping forest to regenerate.

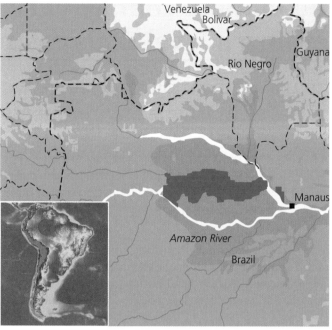

Figure 7.3.1 The location of the Central Amazon Conservation Complex

Be aware that small-scale projects are often more successful than large-scale ones as the people running them have immediate benefits from them.

Sustainable development
Development which improves basic living standards but not at the expense of future generations.

TESTED ◯

4 Identify one advantage and one disadvantage of the Central Amazon Conservation Complex.
5 Identify one advantage and one disadvantage of a small-scale (localised) project.
6 State the main objective of Brazil's Forest Code.

Revision activity

Explain how fig trees help the rainforest to regenerate.

Responses to global warming and climate change

REVISED ◯

Responses to climate change can take two main forms – mitigation and adaptation. Mitigation refers to programmes to try to prevent climate change from happening or to try to reduce the scale of climate change that occurs. Adaptation refers to measures that are taken to manage the impacts of climate change, e.g. preventing coastal erosion, protecting cities against sea level rise.

Mitigation Attempts to prevent climate change from happening.

Adaptation Measures to manage the impacts of climate change.

Individual choices

There are many actions that individuals can do to reduce their own contribution to climate change. These include walking or cycling rather than using a car; eating less meat and dairy products; switching to renewable sources of energy.

The UK's response

Adaptation

There are many ways in which the UK is adapting to climate change. These include improving and increasing flood defences (Figure 7.3.2), improving public transport, and more bus and cycle lanes.

Figure 7.3.2 The Thames Barrier in London

Mitigation

The UK's progress on climate change mitigation is predicted to stall in the 2020s, partly as a result of a lack of new climate policies in recent years, failure to meet afforestation targets and a lack of progress in developing carbon capture and storage (CSS) technology.

Nevertheless, the UK now burns far less coal than it used to. There is potential for more renewable energy, especially wind and solar power.

> **Carbon capture and storage (CCS)** Capturing carbon emissions and then storing the carbon underground in safe rock formations.

Table 7.3.2 UK: Climate change pledges and targets

Paris Agreement	Ratified	Yes
	2030 (unconditional targets)	Greenhouse gas emissions 57% below 1990 levels
	Long-term goal	Net zero greenhouse gas emissions by 2050

Responses to global warming and climate change – China

Mitigation

Table 7.3.3 China: Climate change pledges and targets

Paris Agreement	Ratified	Yes
	2030 (unconditional targets)	Peak CO_2 emissions by 2030
		Non-fossil fuel share 20% in 2030

China has responded in many ways to the threat of global climate change. It has reduced its consumption of coal. This is partly due to changes in China's economy – there is more growth in the service sector and less growth in heavy industries, which can be energy-intensive and use vast amounts of cement and concrete. In 2015, China announced its coal consumption would peak by 2020 and that it would not be building any new coal-fired power stations. China has also invested heavily in wind and solar energy. The Three Gorges Dam makes a significant contribution to the production of renewable energy although the vast amount of concrete used in building the dam would have contributed to global climate change.

China is also providing incentives for buying hybrid vehicles and electric vehicles, as well as enforcing stricter fuel-efficiency standards. China's high-speed trains transport nearly 3 million passengers daily, thereby reducing large numbers of people from using motor vehicles.

> **Hybrid vehicles** Vehicles than can run on electricity as well as fossil fuels.

Adaptation

China has a very varied climate and ecosystems which makes adaptation to climate change complicated. Urban populations, such as Shanghai, are extremely vulnerable to sea level rise, while water scarcity in the north has led to changes in crop productivity, increased flood risk, and more frequent and intense droughts.

Now test yourself

TESTED ◯

7 Outline methods in which individuals could help contribute to reducing climate change.

8 Suggest how China's energy consumption is changing.

9 Comment on the UK's attempts to reduce climate change.

Exam practice

1 The most common greenhouse gas is: (1 mark)
   A carbon dioxide   C water vapour
   B methane   D chlorofluorocarbons

2 Climate change mitigation refers to: (1 mark)
   A measures that deal with the impacts of climate change
   B developments that improve basic living standards without harming future generations
   C attempts to prevent climate change from happening
   D use of technological developments such as HYVs

3 Define the term 'indigenous population'. (1 mark)

4 Suggest why an increase of 2°C could lead to increased risk of death in Africa. (2 marks)

5 Explain **two** potential impacts of rising sea levels. (4 marks)

6 Identify individual choices to deal with global climate change. (4 marks)

7 Explain why the oceans are experiencing acidification. (2 marks)

8 Explain why the sedentarisation of nomads can lead to desertification. (2 marks)

9 To what extent is it possible to manage tropical rainforests sustainably? (6 marks)

10 Examine the main measures to tackle desertification in developed and emerging/developing countries. (12 marks)

Total: 35 marks

Summary

+ There are many types of fragile environments with different characteristics and distributions.
+ There are many causes of desertification and deforestation.
+ Causes of climate change can be natural as well as human.
+ Desertification has social, economic and environmental impacts.
+ Deforestation has social, economic and environmental impacts.
+ Climate change has many negative impacts, e.g. rising sea levels, increased hazards, ecosystem changes,

reduced employment opportunities, changing settlement patterns and health and well-being challenges.
+ Technology can overcome water shortages in fragile environments at risk of desertification.
+ Rainforests can be managed sustainably to reduce the risk of deforestation.
+ Individuals, organisations and governments can take measures to reduce the risk of global warming and climate change.

8 Globalisation and migration

8.1 Globalisation is creating a more connected world

Rise of the global economy

REVISED ⬤

+ Globalisation is a term that describes rapidly increasing global links.
+ Most political borders are not the obstacles they once were so goods, capital, labour and ideas flow more freely across them than ever before.
+ As a result, international trade has increased, encouraging the growth in production of goods and services.
+ The global economy has expanded significantly in recent decades. It was valued at $80 trillion in 2017, up from $41 trillion in 2000.

The development of commodity chains has been an important part of this process. Each stage of manufacturing is usually completed where production costs are lowest. This may involve production in a number of different countries.

Factors encouraging the globalisation of economic activity include:
+ **Reduction of trade barriers:** the barriers to international trade (tariffs, quotas, regulations) are much lower today than in the past.
+ **Foreign direct investment:** Low- and middle-income countries often rely heavily on developed countries to provide investment to develop large businesses and infrastructure. Such foreign direct investment (FDI) has increased rapidly over the last 60 years.
+ **Labour** migration: Fast-growing economies often experience shortages of workers. Potential employees in lower income counties are attracted to the jobs available in richer countries. Labour migration has been at its highest levels ever in the last 20 years.
+ **Advances in transportation:** Major advances in transportation have reduced the geographical barriers separating countries and peoples.
+ **Rapid development of information technologies:** Advances in information technology have affected all aspects of global economic activity. The number of internet users around the world increased from 361 million in 2000 to 4.4 billion in 2019.
+ **Development aid:** Many low-income countries have been able to invest more money in education, health, infrastructure and other aspects of their economies due to financial help from richer nations.

> Globalisation The increasing interdependence of the world economically, culturally and politically.
>
> Commodity chain The stages involved in manufacturing a finished product (commodity) for sale to consumers. These stages may occur in factories in different countries.
>
> International trade The exchange of goods and services between countries.
>
> Tariffs A tax applied by a country on products imported from other countries.
>
> Foreign direct investment (FDI) Investment in physical capital in other countries by transnational corporations.
>
> Migration The movement of people across a specified boundary, national or international, to establish a new permanent place of residence.
>
> Labour migration Migration from one country to another when the main reason is to seek employment.

97

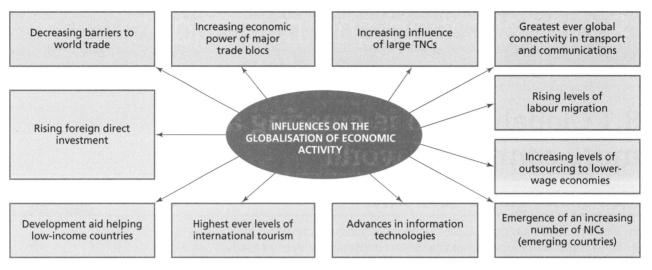

Figure 8.1.1 Factors encouraging the globalisation of economic activity

Now test yourself

1 Define globalisation.
2 List three factors encouraging the globalisation of economic activity.
3 What is foreign direct investment?

Revision activity

Briefly explain the barriers to international trade.

Exam tip

It is easy to think of globalisation as only an economic phenomenon, but it has many other aspects such as cultural and political interactions.

Role of global institutions

REVISED ○

Major global institutions have played an important role in creating a more globalised economy. Their benefits include:

+ pooling expertise from many countries
+ acting as intermediaries between countries in dispute
+ promoting research and development.

Three of the most important global institutions are the World Trade Organization, the International Monetary Fund, and the World Bank. All of these organisations have their supporters and critics!

The World Trade Organization (WTO)

+ The World Trade Organization deals with the rules of world trade.
+ Its main aim is to ensure that world trade flows as freely as possible.
+ The WTO was set up in 1995 with far greater powers than its predecessor (the General Agreement on Tariffs and Trade, GATT) to settle trade disputes.
+ Today average tariffs are only about a tenth of what they were in 1947 when the GATT was formed.

The International Monetary Fund (IMF) and the World Bank

The IMF and the World Bank are United Nations organisations. In more recent times, both organisations have been mainly involved in helping developing countries.

+ A country running seriously short of the foreign currency reserves it needs to maintain its international trade and other commitments can turn to the IMF for help.

- IMF funds come from the contributions of its member countries.
- The IMF is also able to renegotiate the terms of debt on behalf of countries in financial difficulties.
- To prevent the situation reoccurring, the IMF will usually impose conditions relating to how a country runs its economy. Such conditions have proved to be very controversial!

The World Bank provides loans and grants to developing countries. It aims to reduce poverty by increasing economic growth. The World Bank deals mainly with internal investment projects such as building dams and major roads, and promoting health and education.

It often funds projects that would not otherwise be undertaken because:
- The cost is too high for developing countries and they cannot obtain funds elsewhere.
- The benefits are mainly social rather than financial and commercial banks are not interested.

Transnational corporations (TNCs)

Transnational corporations (TNCs) and nation states (countries) are the two main elements of the global economy. The governments of countries individually and collectively (global institutions) set the rules for the global economy, but the bulk of investment is through TNCs.
- TNCs are involved in all economic sectors and have a huge impact on the global economy.
- The hundred largest TNCs represent a significant proportion of total global production.
- Initially manufacturing industry and more recently services have relocated in significant numbers from developed countries to selected developing countries, as TNCs have taken advantage of lower labour costs and other ways to reduce costs.
- It is this process which has resulted in the emergence of an increasing number of newly industrialised countries (emerging countries) since the 1960s. Major examples are China, India and Brazil.

> **Transnational corporation** A firm that owns or controls productive operations in more than one country through foreign direct investment.

Now test yourself

TESTED ◯

4 What is the main aim of the World Trade Organization?
5 Which international organisation would a country approach if it was seriously short of foreign currency reserves?
6 Define transnational corporations.

Revision activity

List the benefits of major global institutions. Name a global institution other than the ones discussed in this section.

Movements of people

REVISED ◯

Push and pull factors

International migrations start from a country of origin and are completed at a country of destination. Push and pull factors (Figure 8.1.2) encourage people to migrate. They can be divided into the following categories:
- economic
- social
- political
- environmental.

For example, a high level of unemployment is a major push factor. In contrast, an important pull factor is often much higher wages in another country. The nature of push and pull factors varies from country to country (and from person to person) and changes over time.

> **Push and pull factors** Push factors are the negative factors at the place of origin; pull factors are the positive factors at the destination.

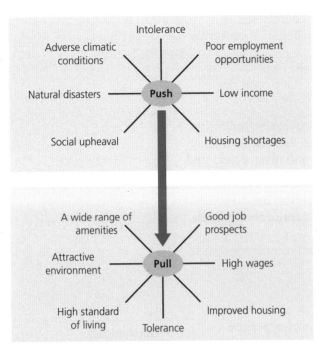

Figure 8.1.2 Push and pull factors

Voluntary and forced migrations

There is a clear distinction between voluntary and forced migrations (Figure 8.1.3). Violent conflicts have displaced huge numbers of people in recent decades. Not all have crossed international frontiers to merit the term refugee movements. Instead many are internally displaced people. The long-lasting conflict in Syria has produced large numbers of both refugees and internally displaced people.

In most countries there are no legal restrictions on internal migration. Thus, the main constraints are distance and cost. In contrast, immigration laws present the major barrier in international migration. Most countries favour immigration applications from highly skilled people.

Voluntary migration When individuals have a free choice about whether to migrate or not.

Forced migration When people are made to move against their will due to human or environmental factors.

Internally displaced people People forced to flee their homes due to human or environmental factors, but who remain in the same country.

Refugees People forced to flee their homes due to human or environmental factors and who cross an international border into another country.

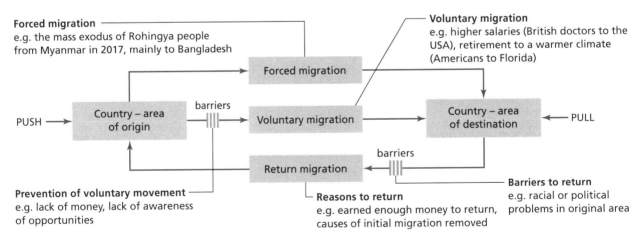

Figure 8.1.3 Types of migration and barriers to migration

Migration trends

Increasing rates of international migration and tourism are major elements of globalisation.

+ Almost 272 million people lived outside the country of their birth in 2019, up from 102 million in 1980. This is about 3.5% of the world's population.
+ Two-thirds of this total are labour migrants.
+ Of international migrants, 52% are male, 48% are female.
+ Most international migrants (74%) are of working age (20–64 years).
+ Globalisation has led to a) an increased awareness of opportunities in other countries; and b) an increasingly high level of potential mobility.

Much international migration is economic – people seeking employment in other countries. However, people also move for other reasons:

+ Older people retiring to countries with warmer climates, e.g. from northern to southern Europe.
+ Professional sports players moving to clubs in other countries which offer higher wages and a larger media profile.
+ Medical reasons – medical professionals in demand internationally and patients migrating for health reasons.

Sometimes tourism is considered to be a form of migration. However, most academics see tourism as a separate entity as it is mainly very short term in nature. In 2012 international tourist arrivals exceeded 1 billion for the first time ever. The World Tourism Organization (UNWTO) forecasts an increase to 1.8 billion in 2030.

Internal population movements

+ Population movement within countries is at a much higher level than movement between countries.
+ Much internal migration is from a) poorer to richer regions and b) rural to urban areas, to seek employment and a higher standard of living.
+ Developed countries had their period of high rural-to-urban migration in the nineteenth century and early part of the twentieth century. Developing and emerging countries have been undergoing high rural-to-urban migration since the 1950s.

> **Rural-to-urban migration**
> The movement of large numbers of people from the countryside to towns and cities.

> **Revision activity**
> Study Figure 8.1.2. Place the individual push and pull factors into the four categories identified in the first paragraph of this section.

> **Exam tip**
> Remember that forced migration is not just the result of armed conflict, but can also occur due to environmental factors such as volcanic eruptions and desertification.

Now test yourself  TESTED ○

7 Define migration.
8 Explain the terms 'origin' and 'destination'.
9 What proportion of the world's population lives outside the country of their birth?

8.2 The impacts of globalisation

Impacts of globalisation on different groups of people

REVISED ○

+ The impacts of globalisation vary at different scales (Table 8.2.1).
+ Some countries have benefited more than others, particularly developed nations and those which have transformed themselves from developing to emerging economies.
+ Within individual countries, some industries may benefit from greater access to foreign markets, while other industries may struggle against intense international competition.
+ Highly skilled workers have benefited much more from globalisation than those with limited skills.
+ The gender gap within individual countries is generally lower in more globalised countries.

Table 8.2.1 Examples of the impacts of globalisation at global, national and local scales

Global	National	Local
The growing power of TNCs and global brands	Concerns about loss of sovereignty to regional and international organisations	Small local businesses often find it difficult to compete with major global companies
The development of an increasing number of emerging economies	Increased cultural diversity from international migration	Closure of a TNC branch plant can cause high local unemployment
Development of a hierarchy of global cities	Higher levels of incoming and outgoing international tourism	The populations of many local communities have become more multi-cultural
The emergence of powerful trade blocs	TNCs employing an increasing share of the workforce	Families now more likely to be spread over different countries due to increased international migration
Environmental degradation caused by increasing economic activity	Increasing incidences of trans-boundary pollution	The development of 'ethnic villages' in large urban areas

The benefits and costs to countries hosting TNCs

There are many advantages for companies in working at the global scale:
+ Sourcing raw materials and components on a global basis reduces costs.
+ TNCs can seek out the lowest cost locations for labour.
+ Selling goods and services to a global market allows TNCs to achieve large economies of scale.
+ Global marketing helps to establish brands with huge appeal all around the world.

> **Economies of scale** The reduction in unit cost as the scale of production increases.
>
> **Brand** A clearly recognisable name and/or symbol intended to identify a product or producer.

The way TNCs operate has come under much criticism. This relates not just to the impact on poorer economies, but also to the impact on the countries of origin of TNCs. Table 8.2.2 summarises the potential costs and benefits of Nike to the USA, its headquarters' location, and to Vietnam, where many subcontracted Nike factories operate.

Table 8.2.2 The costs and benefits of TNCs

Country	Possible advantages	Possible disadvantages
USA: headquarters	Positive employment impact and stimulus to the development of high-level skills in design, marketing and development in Beaverton, Oregon; direct and indirect contribution to local and national tax base	Another US firm that does not manufacture in its own country — indirect loss of jobs and the negative impact on balance of payments as footwear is imported; trade unions complain of an uneven playing field because of the big contrast in working conditions between developing and developed countries
Vietnam: outsourcing	Creates substantial employment in Vietnam; pays higher wages than local companies; improves the skills base of the local population; the success of a global brand may attract other TNCs to Vietnam, setting off the process of cumulative causation; exports are a positive contribution to the balance of payments; sets new standards for indigenous companies; contribution to local tax base helps pay for improvements to infrastructure	Concerns over the exploitation of cheap labour and poor working conditions; allegations of the use of child labour; company image and advertising may help to undermine national culture; concerns about the political influence of large TNCs; the knowledge that investment could be transferred quickly to lower-cost locations

Now test yourself

TESTED

1 State two advantages for companies in working at the global scale.
2 Name two products which are considered to be global brands.
3 Give two possible disadvantages of TNCs to developing countries.

Revision activity

Study Figure 8.2.1. Write paragraphs to elaborate on three of the impacts presented.

Exam tip

Remember that the activities of TNCs can, at times, be very controversial. Try to look for positive as well as negative impacts, and keep in mind that some operate in a much more socially responsible way than others.

Impacts of migration

REVISED

Voluntary migration

Figure 8.2.1 shows some of the possible impacts of voluntary international migration on:

- countries of origin
- countries of destination
- migrants themselves.

Many of these factors are also relevant to internal migration in terms of regions of origin and destination.

The impact of international migration		
Impact on countries of origin	**Impact on countries of destination**	**Impact on migrants themselves**
Positive • Remittances are a major source of income in some countries. • Emigration can ease the levels of unemployment and underemployment. • Reduces pressure on health and education services and on housing.	• Increase in the pool of available labour may reduce the cost of labour to businesses and help reduce inflation. • Increasing cultural diversity can enrich receiving communities. • An influx of young migrants can reduce the rate of population ageing.	• Wages are higher than in the country of origin. • There is a wider choice of job opportunities. • They have the ability to support family members in the country of origin through remittances.
Negative • Loss of young adult workers who may have vital skills, e.g. doctors, nurses, teachers, engineers (the 'brain-drain' effect). • An ageing population in communities with a large outflow of (young) migrants. • Agricultural output may suffer if the labour force falls below a certain level.	• Migrants may be perceived as taking jobs from people in the long-established population. • Increased pressure on housing stock and on services such as health and education. • A significant change in the ethnic balance of a country or region may cause tension.	• The financial cost of migration can be high. • Migration means separation from family and friends in the country of origin. • There may be problems settling into a new culture (assimilation).

Figure 8.2.1 Matrix showing the impact of migration

Remittances are a major economic growth factor in developing countries.

+ Remittances to developing countries totalled $429 billion in 2016, exceeding the amount of official development aid.
+ Remittances can account for over 20% of annual GDP in some developing countries.
+ The top recipients of remittances in 2016 were: India, China, the Philippines and Mexico.

> **Remittances** Money sent back by migrants to their families in their home communities.

Each receiving country has its own sources of immigrants. This is a result of historical, economic and geographical relationships.

Forced migration

+ The number of forcibly displaced people worldwide stood at 79.5 million at the end of 2019. This included 26 million refugees.
+ Worldwide, 85% of refugees are hosted in developing countries.
+ Many are housed at very high densities in huge refugee camps. Apart from immediate problems concerning overcrowding, sanitation and the disposal of waste, long-term environmental damage may result from deforestation and soil degradation.
+ The high cost of refugee camps is financed by host countries and international charities.

Rural–urban migration

Rural–urban migration has led to the massive expansion of many urban areas in developing countries. It has been a major factor in the development of megacities, resulting in large areas of slum housing where problems include overcrowding, sanitation, lack of essential infrastructure, and high unemployment and underemployment. On the other hand, a large pool of labour may be seen as an attraction for TNCs.

Very high levels of migration from rural areas can lead to rural depopulation. As such migration is usually dominated by young adults, the rural areas affected exhibit ageing populations. These trends often lead to a decline in rural service provision (Figure 8.2.2). In developing countries in particular, agricultural production may fall if too few farm workers are left in rural areas.

> **Rural depopulation** The absolute decline in the population of a rural area, frequently caused by out-migration.

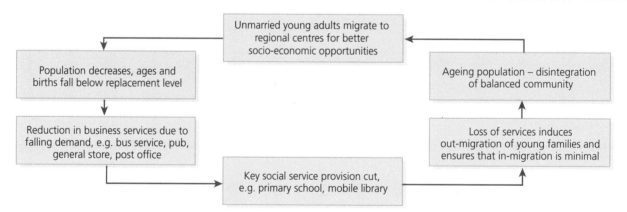

Figure 8.2.2 Model of rural depopulation

4 Define remittances.

5 How many people around the world are forcibly displaced?

6 What is rural depopulation?

Exam tip

Rural-to-urban migrants are not necessarily the poorest people from rural areas, particularly when attractive opportunities are available in urban areas.

Impacts of the growth of tourism

REVISED

Over the last fifty years, tourism has developed into a major global industry which is still expanding rapidly.

The economic impact

Supporters of the development potential of tourism say that:

+ Tourism brings in valuable foreign currency and benefits other sectors of the economy by creating a multiplier effect (see Figure 8.2.3).
+ It provides considerable tax revenues.
+ By providing employment in rural areas, it can help to reduce rural–urban migration.
+ It can create openings for small businesses and support many jobs in the informal sector.

Tourism Travel away from the home environment for: a) leisure, recreation and holidays, b) to visit friends and relations (VFR), and c) for business and professional reasons.

Multiplier effect The idea that an initial investment causes money to circulate in the economy, bringing a series of economic benefits over time.

Economic leakages The part of the money a tourist pays for a foreign holiday that does not benefit the destination country because it goes elsewhere.

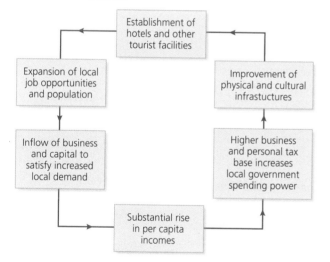

Figure 8.2.3 The multiplier effect of tourism

However, critics say that the value of tourism is often overrated because:

+ Economic leakages (Figure 8.2.4) are high.
+ Most local jobs created are menial, low paid and seasonal.
+ Money borrowed to invest in the necessary infrastructure for tourism increases the national debt.
+ At some destinations, tourists spend most of their money in their hotels with minimum benefit to the wider community.
+ Tourism might not be the best use for local resources which could, in the future, be put to better use by a different economic sector.

Figure 8.2.4 Diagram of economic leakages

The social and cultural impact of tourism

Table 8.2.3 The social and cultural impact of tourism

Social/cultural disadvantages	Social/cultural advantages
✦ Loss of locally owned land ✦ Abandonment of traditional values ✦ Displacement of people ✦ Traditional community structures may be weakened ✦ Abuse of human rights ✦ Increasing availability of alcohol and drugs ✦ Crime and prostitution, sometimes involving children ✦ Visitor congestion at key locations ✦ Denying local people access to beaches ✦ Loss of housing for local people as more visitors buy second homes	✦ Can increase the range of social facilities for local people ✦ Can lead to greater understanding between people of different cultures ✦ Visiting ancient sites can develop a greater appreciation of the historical legacy of host countries ✦ Can help develop foreign language skills in host communities ✦ May encourage migration to major tourist-generating countries ✦ Major international events such as the Olympic Games can have a very positive global impact

Environmental impact

Tourism has reached such a large scale in so many parts of the world that it can only continue with careful management. However, sustainable tourism strategies have been more successful in some areas than others.

Examples of the negative environmental impact of tourism include:
+ the destruction of coral reefs by tourist boats and water sports
+ disturbance of wildlife and flora
+ congestion and overcrowding at 'honeypot' locations
+ increased greenhouse gas emissions
+ new golf courses reducing scarce water supplies for local communities.

The environmental impact of tourism is not always negative. Landscaping and sensitive improvements to the built environment have improved the overall quality of some areas. On a larger scale, tourism revenues can fund the designation and management of protected areas such as national parks.

Sustainable tourism
Tourism organised in such a way that its level can be sustained in the future without creating irreparable environmental, social and economic damage to the receiving area.

Revision activity
Expand briefly on two of the economic leakages given in Figure 8.2.6.

Exam tip
It is easy to fall into the trap of only seeing the disadvantages of the cultural impact of tourism. It is always important to consider the other side of the coin, even if you can only come up with a few points.

Now test yourself   TESTED ◯

7   What is a multiplier effect?
8   Define economic leakages.
9   Give three social/cultural disadvantages of the growth of tourism.

Answers to Now Test Yourself and Exam Practice questions at **www.hoddereducation.co.uk/myrevisionnotesdownload**

8.3 Responses to increased migration and tourism

Geopolitical relationships between countries

+ International trade, migration and tourism all involve large-scale interactions between countries.
+ Never before have the geopolitical relations between countries been so important.
+ Individual countries try to ensure that the benefits of such movements outweigh the costs.
+ To achieve this objective, countries try to manage these movements to varying degrees. This often requires political agreement between countries.
+ Countries at a higher level of development have more power to develop positive outcomes than poorer nations.

> **Geopolitics** Political relations between countries, particularly relating to claims and disputes concerning borders, territories and resources.
>
> **Trade bloc** A group of countries that share trade agreements between each other.
>
> **Terms of trade** The price of a country's exports relative to the price of its imports.

Trade

The number of regional trade agreements (trade blocs) has increased significantly over the last forty years. The most notable of these are the European Union (EU), NAFTA in North America, ASEAN in Asia, and Mercosur in Latin America. Most trade blocs allow free trade between member countries, while trade with countries outside of the trade bloc is subject to tariffs. Countries within trade blocs differ in size and economic power. It is not surprising that the most powerful countries tend to dictate policies in most trade blocs.

The most vital element in the trade of any country is the terms on which it takes place (the terms of trade). Many developing countries feel that the WTO is dominated by developed countries and not enough has been done to help the trading position of the world's poorest countries.

Migration

The International Organization for Migration (IOM) is the leading international organisation for migration. It works to:
+ advance understanding of migration issues
+ assist in the challenges of migration management.

The IOM works with other international organisations to try to broker disputes between countries about population movements, particularly with regard to: a) refugees and b) illegal immigration. In general, developing countries want their populations to have greater migration access to developed countries. In contrast, developed countries have tended to tighten controls on immigration. Within the European Union (EU), there is free movement of people between member countries.

Tourism

+ The increasing economic importance of tourism can have a positive effect on the relationship between countries that may have previously only had limited or frosty relations.
+ For countries that have been relatively 'closed' to the outside world, tourism can be important in the process of engaging more positively in the wider world.
+ International tourism can help promote understanding of different cultures.
+ Tourism can be a precursor to a permanent migration stream.

Now test yourself

TESTED ◯

1 Define geopolitics.
2 What is a trade bloc?
3 Name the leading global institution for migration.

Revision activity

Name four trade blocs from four different continents.

Exam tip

Trade blocs can vary in their degree of economic and political integration. For example, the EU is a much more complex organisation than NAFTA.

Different approaches to the management of migration

REVISED ◯

International migration is a major global issue. Today, few countries favour a large influx of migrants for a variety of issues. Within individual countries attitudes to immigration can vary considerably. Attitudes tend to harden when economies are going through difficult periods and become more relaxed when economic conditions are good.

100 years of immigration to the US
Persons obtaining lawful permanent resident status, 1900–2009, US Department of Homeland Security

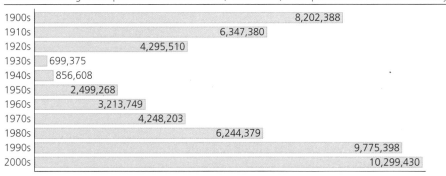

Decade	Persons
1900s	8,202,388
1910s	6,347,380
1920s	4,295,510
1930s	699,375
1940s	856,608
1950s	2,499,268
1960s	3,213,749
1970s	4,248,203
1980s	6,244,379
1990s	9,775,398
2000s	10,299,430

Figure 8.3.1 Immigration to the USA by decade 1900s–2000s

There have been many changes to immigration regulations in the USA as government concerns about immigration have shifted over time. Some of the most notable changes have included the following:

✦ Beginning in 1875, a series of restrictions on immigration were enacted, e.g. limiting the rising number of Asian immigrants.
✦ By the early 1900s, the main migration flows were moving from the countries of northern and western Europe to the nations of southern and eastern Europe. In 1924 a system of 'national origins quotas' was introduced which operated with only slight modification until 1965. This legislation aimed to: a) greatly reduce immigration and b) preserve the ethnic balance of the USA which existed at the 1920 census. It offered the largest quotas of entry permits to British, Irish and German immigrants (70% in total).
✦ The racist overtones of the 1924 system, resulting in considerable internal and international opposition, led to its abolition in 1965. The 1965 Act set an annual limit of 120,000 immigrants from the Western Hemisphere (the Americas) and 170,000 from the Eastern Hemisphere. People from every country now had an equal chance of acceptance. However, resulting immigration exceeded this level because relatives of US citizens were admitted without limitation. Europe, the previous major source region, was overtaken in 1970 by the rest of the Americas and by Asia, a trend that has continued to the present day.
✦ The 1986 Immigration Reform and Control Act granted legal status to millions of unauthorised immigrants, mainly from Latin America, who met certain conditions. The Act imposed sanctions on employers who hired illegal immigrants.

Government Act A law enacted by a legislature, e.g. a parliament.

Answers to Now Test Yourself and Exam Practice questions at **www.hoddereducation.co.uk/myrevisionnotesdownloa**

- The Immigration Act of 1990 raised immigration quotas by 40%.
- Later laws in 1996, 2002 and 2006 attempted to address concerns relating to terrorism and illegal immigration.
- In 2012, executive action by President Obama allowed young adults, who had been brought to the USA illegally, to apply for deportation relief and a work permit. In 2014, a new programme offered similar benefits to some unauthorised immigrant parents of children born in the USA.
- There has been an increase in the removal of illegal immigrants since President Trump took office in 2017. The President also announced that a wall would be built along the US–Mexican border to reduce illegal immigration.

One of the largest international migration streams in the world over the past fifty years has been from Mexico to the USA. This migration has formed the largest immigrant community in the world. It has been primarily a labour migration, the result of:
- much higher average incomes in the USA
- lower unemployment rates in the USA
- the faster growth of the labour force in Mexico
- the much better quality of life in the USA.

Most migration from Mexico has taken place since the 1970s. However, previous surges occurred in the 1920s and 1950s, when the American government allowed the recruitment of Mexican workers as guest workers.

> **Guest workers** A foreigner who is permitted to work in a country on a temporary basis, e.g. a farm labourer.

> **Revision activity**
>
> Study Figure 8.3.1. Describe how the volume of immigration has changed from the early 1900s.

> **Exam tip**
>
> Students at times make very generalised statements about attitudes to immigration. Try to be as specific as possible. For example, many employers favour a high rate of immigration because it increases the potential pool of labour, while trade unions sometimes oppose high immigration because this can keep wage rates lower than they would otherwise be.

Now test yourself TESTED ◯

4 What were the main changes brought about by the 1965 Immigration Act?

5 Give two reasons for the high rate of migration from Mexico to the USA.

6 Define a guest worker.

Different approaches to sustainable tourism

REVISED ◯

Approaches to sustainable tourism in Portugal

Tourism is of considerable importance to national income and employment in Portugal. The development of sustainable tourism in the country has revolved around:
- Establishing a system of protected areas – national parks, nature parks, nature reserves, protected landscapes, natural monuments, and private protected areas.
- Reducing the environmental impact of tourism in different types of destination all over the country.

In 2017, the Portuguese government launched its 'Tourism Strategy 2027' with sustainability as its main priority. This programme was based on the UN's Sustainable Development Goals. Key elements included:
- Setting objectives for each of the three pillars of sustainable development (economic, social, environmental).
- Viewing tourism as a vehicle for promoting balanced development in Portugal by decentralising tourism demand to less developed regions throughout the year.
- Adding value to local communities.

A 'sustainable tourism indicators system' was developed to implement Tourism Strategy 2027. This:
- enables tourism policy evaluation
- provides the private sector with instruments for making strategic decisions.

> **Sustainable Development Goals** UN plans to make the world a better, fairer place for all people and to improve the health of the planet.

A wide range of indicators cover environmental, economic and social objectives. Table 8.3.1 gives some examples.

Table 8.3.1 Portugal's Sustainable Tourism Indicators System (STIS)

Environmental	Economic	Social
Quality of water in bathing areas	Proportion of jobs which are seasonal	Accessibility for guests with special needs
Energy consumption and emissions in tourism	Hotels' etc. use of local suppliers	Tourism density
Training in sustainable practices	Occupancy rate	Proportion of tourists returning to Portugal

Sustainable tourism in Ecuador

International tourism is a large source of income in Ecuador. The number of visitors has steadily increased, both to the mainland and to the Galapagos Islands where Darwin conducted research on evolution. The majority of tourists are drawn to Ecuador by its great diversity of flora and fauna. Much of the country is protected by national parks and nature reserves.

Case study

Galapagos Islands at risk

The Galapagos Marine Reserve is one of the largest marine protection areas in the world. In 2007, the government of Ecuador declared the Galapagos Islands at risk because of:
+ the increasing impact of tourism
+ illegal fishing
+ the problem of invasive species.

WWF (World Wide Fund for Nature), along with a number of other NGOs, has worked with the Ecuadorian government to strengthen conservation efforts. For example, WWF has:
+ supported the Galapagos National Park to improve its control and surveillance system

+ promoted an artisanal fishing culture which aims to maximise catches while minimising environmental impact.

Conservation measures have evolved over time. The current focus is on:
+ ecosystem restoration
+ tackling invasive species
+ monitoring indicator species
+ marine management
+ supporting cutting edge research
+ species-specific projects.

Case study

Ecotourism in the Amazon

Another important geographical focus for ecotourism has been in the Amazon rainforest around Tena. The ecotourism schemes in the region are usually run by small groups of indigenous Quichua Indians. The indigenous movement in Ecuador is one of the strongest in South America. Ecotourism provides an important alternative source of income to local communities and is an extra incentive to conserve natural resources. Yasuni National Park is said to have the highest density of wildlife on Earth!

Ecotourism A specialised form of tourism where people experience relatively untouched natural environments such as coral reefs, tropical forests and remote mountain areas, and ensure that their presence does no further damage to these environments.

Now test yourself

TESTED

7 Define sustainable tourism.
8 State two objectives of Portugal's Tourism Strategy 2027.
9 Give two ways in which WWF has helped to promote conservation in the Galapagos Islands.

Revision activity

With reference to an atlas, describe the location of the Galapagos Islands.

Exam tip

Developing an effective sustainable tourism policy requires good organisation, commitment and investment.

1 The major global organisation dealing with world trade is the: (1 mark)
   A IMF
   B WTO
   C WHO
   D UNHCR

2 a) What is globalisation?   (2 marks)
   b) What is the role of transnational corporations in globalisation?   (3 marks)

3 a) There is a clear distinction between voluntary and which other
      type of migration?   (1 mark)
   b) What is the difference between push and pull factors?   (2 marks)
   c) Give two reasons for migration other than to find work.   (2 marks)

4 a) What is a trade bloc?   (2 marks)
   b) Give two examples of a trade bloc.   (2 marks)
   c) Give two examples of barriers to trade.   (2 marks)

5 Examine the major social issues associated with the growth
   of tourism in developing countries.   (6 marks)

6 Discuss the approaches to sustainable tourism in a named country.   (12 marks)

Total: 35 marks

Summary

+ Globalisation is creating a more interconnected and interdependent world, as exemplified by the growth of world trade, foreign direct investment, labour migration and the development of commodity chains.
+ Major advances in information technologies and transportation have been essential for globalisation to develop at such a rapid pace.
+ Major global institutions such as the WTO, the IMF and the World Bank, along with TNCs, have played very important roles in creating a more globalised economy.
+ Migration, with its push and pull factors, and tourism are significant aspects of the global economy.
+ The impacts of globalisation vary on a global scale, as exemplified by the benefits and costs to countries hosting TNCs.

+ Migration, in its various forms, has impacts on different groups of people.
+ The growth of tourism has positive and negative impacts on the environment, economy and people of destination areas.
+ Geopolitical relationships are important in managing trade, migration and tourism.
+ Approaches to the management of migration often change over time within individual countries.
+ Approaches to sustainable tourism can vary between countries, particularly between those at different stages of economic development.

9.1 Defining and measuring human welfare

Different ways of defining development

Development, or improvement in the quality of life, is a wide-ranging concept. It includes increasing wealth, but also involves other important aspects of our lives (Figure 9.1.1). For example, development occurs in a low-income country when:

+ the electricity grid extends outwards from the main urban areas to rural areas
+ levels of literacy improve throughout the country.

The United Nations Development Programme (UNDP) states that:

+ development is about creating opportunities and giving people the freedom to make choices
+ development must balance social, economic and environmental sustainability.

Thus, development is a process of change for the better. It is not always a continuous process as it can be interrupted by major national and global events such as war, famine and economic recession.

> **Development** The use of resources to improve the quality of life in a country.
>
> **United Nations Development Programme [UNDP]** The UN's development agency working to eradicate poverty. It plays a crucial role in helping countries achieve the Sustainable Development Goals.

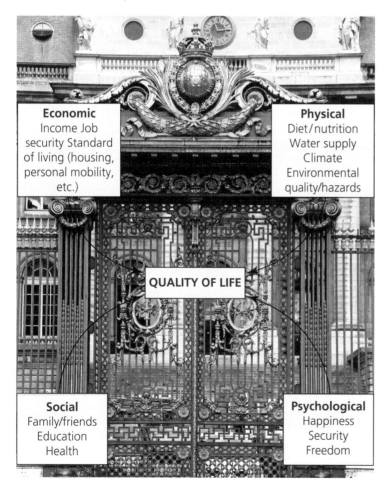

Economic
Income Job security Standard of living (housing, personal mobility, etc.)

Physical
Diet/nutrition
Water supply
Climate
Environmental quality/hazards

QUALITY OF LIFE

Social
Family/friends
Education
Health

Psychological
Happiness
Security
Freedom

Figure 9.1.1 Factors comprising the quality of life

Individual strands of development can be grouped into categories such as in Table 9.1.1.

Table 9.1.1 Categories of development

Category	Examples
Social	Achieving higher levels of literacy; improving life expectancy
Economic	Raising productivity; increasing incomes for all levels of society
Environmental	Reducing air, water and land pollution; conserving biodiversity
Political	Improving democratic institutions; adhering to new international human rights obligations

Revision activity

Look at Table 9.1.1. Try to add an additional example for each of the four categories.

The UN's Sustainable Development Goals (SDGs) were adopted by UN Member States in 2015. The overall objectives were to: a) end poverty; b) protect the planet; and c) ensure all people enjoy peace and prosperity. The 17 SDGs are 'integrated' – recognising that action in one area will affect outcomes in others. The hope was that all these aims would be achieved by 2030.

Exam tip

It is very important to understand the difference between economic growth and development. The former is an increase in gross domestic product (GDP), while development is a more wide-ranging concept concerning many more aspects of the quality of life.

Now test yourself   TESTED ◯

1 What is development?

2 Give two examples of development in a low-income country.

3 The UNDP stresses the importance of a balance between which three types of development?

Factors affecting development and human welfare

REVISED ◯

A range of factors contribute to the development and human welfare of a country. These include economic, social, cultural and technological factors.

Stages of development

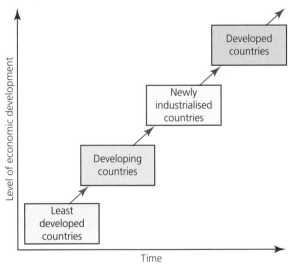

Figure 9.1.2 Stages of development

A reasonable division of the world in terms of economic development is shown in Figure 9.1.2.

The least developed countries (LDCs) are the poorest of the developing countries. Their problems are often made worse by major geographical handicaps. LDCs are often primary product dependent. In 2019, the UN recognised 46 countries as LDCs. Of these, 31 are in Africa.

Least developed countries (LDCs) The poorest of the developing countries. They have major economic, institutional and human resource problems.

Primary product dependent When a country relies on one or a small number of primary products for most of its export earnings.

113

A significant number of LDCs and developing countries have problems with food and water security. In 2018:

+ more than 700 million people were exposed to severe levels of food insecurity
+ 785 million people were living without access to safe water.

Newly industrialised countries (emerging economies) are nations that have moved up the development ladder, having previously been considered developing countries. The first countries to become newly industrialised countries (in the 1960s) were South Korea, Singapore, Taiwan and Hong Kong. The media referred to them as the 'four Asian tigers'. The success of these four countries provided a model for others to follow such as Malaysia, Brazil, China and India.

Explaining the development gap

Reasons for variations in development between countries include:

Physical geography:

+ Landlocked countries have generally developed more slowly than coastal nations.
+ Small island countries face considerable disadvantages in development.
+ Tropical countries have grown more slowly than those in temperate latitudes.
+ A generous allocation of natural resources has spurred economic growth in a number of countries, but there are examples where resource wealth has been wasted.

Economic policies:

+ Open economies which welcomed and encouraged foreign investment have developed faster than closed economies.
+ Fast-growing countries tend to have high rates of saving and low spending relative to gross domestic product (GDP).
+ Institutional quality in terms of good government, law and order, and lack of corruption generally results in high rates of growth.

Demography:

+ Progress through demographic transition is a significant factor, with the highest rates of economic growth experienced by those nations where the birth rate has fallen the most.

Figure 9.1.3 combines a range of factors to explain differences in development. For example, in diagram (a) Brazil would satisfy all three criteria. In contrast, countries such as Haiti and Niger would be affected by all three of the negative factors in diagram (b).

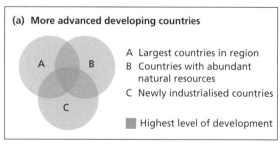

(a) **More advanced developing countries**

A Largest countries in region
B Countries with abundant natural resources
C Newly industrialised countries

▪ Highest level of development

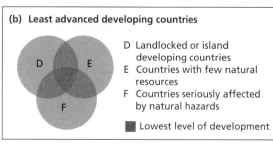

(b) **Least advanced developing countries**

D Landlocked or island developing countries
E Countries with few natural resources
F Countries seriously affected by natural hazards

▪ Lowest level of development

Figure 9.1.3 Fast and slow development in developing countries

---

Food security When the whole population of a country, at all times, has access to sufficient, safe and nutritious food.

Water security The capacity of a population to safeguard sustainable access to adequate quantities of acceptable quality water.

Newly industrialised countries (emerging economies) Nations that have undergone rapid and successful industrialisation/ economic growth since the 1960s.

Gross domestic product (GDP) The total value of goods and services produced within a country in a year.

Revision activity

Write a paragraph to explain Figure 9.1.3.

Exam tip

The evidence seems to suggest that it has been more difficult to progress from LDC status to 'developing' than from the latter to the level of newly industrialised country (emerging economy).

Answers to Now Test Yourself and Exam Practice questions at **www.hoddereducation.co.uk/myrevisionnotesdownloa**

4 In which continent are the majority of LDCs located?

5 Define a) food security; and b) water security.

6 Give two factors of physical geography that can inhibit economic development.

Development indices

REVISED ○

GDP per capita

Gross domestic product (GDP) per capita is the total value of goods and services produced within a country in a year (GDP) divided by the number of people in the country. Per capita figures allow comparisons to be made between countries with big differences in population size. Economic growth (GDP growth) refers to an increase in real GDP (%) after inflation is taken into account.

While GDP per capita is a useful measure, it does not tell us anything about:
+ wealth distribution within a country
+ the ways in which wealth is used.

The human development index

The human development index (HDI) was devised by the United Nations in 1990. The current index combines four indicators of development (Figure 9.1.4):
+ life expectancy at birth
+ mean years of schooling for adults aged 25 years
+ expected years of schooling for children of school-entering age
+ gross national income (GNI) per capita (PPP $).

Gross domestic product (GDP) per capita The total GDP of a country is divided by the total population.

Economic growth An increase in real GDP (%) after inflation is taken into account.

Human development index (HDI) The composite UN measure of the disparities between countries using indicators of health, education and income.

PPP Purchasing power parity, which takes into account variations in the cost of living between countries.

Corruption Dishonest or illegal behaviour by people in powerful positions, typically involving bribery.

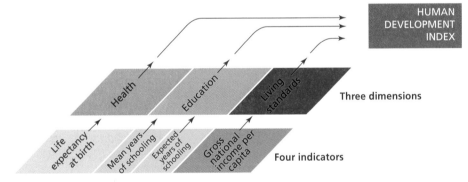

Figure 9.1.4 The components of the human development index

The HDI divides the countries of the world into four groups:
+ Very high human development (e.g. Canada, UK, Japan, Australia).
+ High human development (e.g. Russia, Brazil, Saudi Arabia, Turkey).
+ Medium human development (e.g. China, India, South Africa, Indonesia).
+ Low human development (e.g. Pakistan, Nigeria, Tanzania, Papua New Guinea).

Indices of political corruption

The Corruption Perceptions Index 2019 ranked 180 countries for perceived levels of public sector corruption.
+ Low levels of corruption – Canada, Singapore, New Zealand and Australia, and north-west Europe.
+ High levels of corruption – Russia, much of the Middle East, Africa and Latin America.

Corruption has a large negative impact on development. Among other things, it:

+ takes money from public finances
+ creates distrust between the general public and the government.

Measures of inequality

The Gini coefficient is a technique frequently used to show the extent of income inequality within countries.

+ It is defined as a ratio with values between 0 and 1.0.
+ A low value indicates a more equal income distribution, while a high value shows more unequal income distribution.
+ In general, more affluent countries have a lower income gap than developing countries.
+ Southern Africa and South America show up clearly as world regions of very high income inequality. Europe is the world region with the lowest income inequality.

> **Gini coefficient** A statistical technique used to show the extent of income inequality in a country.

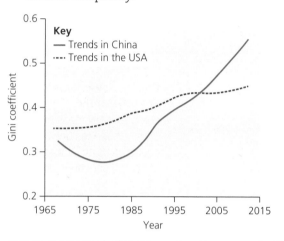

Figure 9.1.5 Graph showing regional inequality in China and USA

Figure 9.1.5 shows how regional inequality has changed in China and the USA since the mid-1960s. Since the early 2000s, regional disparities in China have become greater than in the USA and the gap has been getting wider.

> **Exam tip**
>
> It is important to stress that every individual indicator of development has both merits and limitations. Composite measures such as the HDI try to overcome this situation. However, there is always debate about which indicators to include in a composite measure.

Now test yourself

TESTED ◯

7 Define GDP per capita.
8 Give two reasons to explain why corruption hinders development.
9 Name two parts of the world experiencing very high income inequality.

> **Revision activity**
>
> List the indicators used in the human development index (HDI).

9.2 Uneven development

Uneven development within and between countries

REVISED ◯

Uneven development between countries

Figure 9.2.1 shows the global variation in GDP per capita for 2018.

+ The world regions of highest development are: the USA and Canada; much of Europe; Japan; Singapore; Australia and New Zealand; the oil-rich countries of the Middle East.
+ The regions of lowest development are: Africa; western, southern and south-eastern Asia; parts of Latin America.

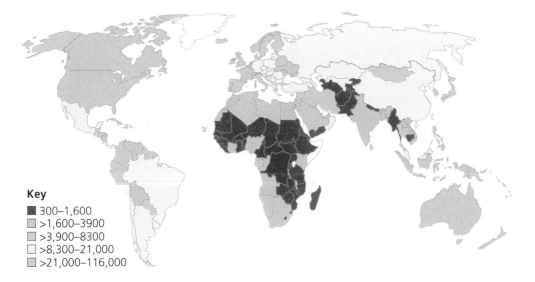

Key
■ 300–1,600
■ >1,600–3900
■ >3,900–8300
□ >8,300–21,000
■ >21,000–116,000

Figure 9.2.1 World map showing GDP per capita, 2018 (US$)

Many of the reasons for the development gap have already been covered, but other factors that also merit attention include those listed below.

Colonialism and neo-colonialism
✦ Colonialism occurred mainly in the eighteenth and nineteenth centuries, as European powers such as the UK, France and Spain expanded their territories around the world.
✦ They exploited their colonies for economic gain.
✦ In the modern world, the term 'neo-colonialism' is used to describe how rich countries can still dominate poorer countries in an economic and political sense.
✦ Dependency theory was developed to help explain the impact of these processes.

Political and economic policies
Open economies such as the UK and Singapore that encourage foreign investment have developed faster than countries that have much greater controls on foreign investment. Large-scale investment can create significant economic multiplier effects.

Social investment
Countries that have prioritised investment in education and health have generally developed at a faster rate than nations that have invested less in these sectors. High standards of education and health are vital for sustainable development.

Uneven development within countries
The model of cumulative causation helps to explain regional disparities. Figure 9.2.2 is a simplified version of the model. There are three stages:
✦ the pre-industrial stage when regional differences are small
✦ a period of rapid economic growth with increasing regional economic divergence
✦ a stage of regional economic convergence.

Development gap The difference in income and the quality of life in general between the richest and poorest countries in the world.

Colonialism The establishment of colonies in one or a number of territories by people from another territory.

Neo-colonialism The dominance of poor countries by rich countries, not by direct political control (as in colonialism), but by economic power and cultural influence.

Dependency theory A theory which blames the relative underdevelopment of the developing world on exploitation by the developed world, first through colonialism and then by neo-colonialism.

Open economy A country with few investment and trade barriers, encouraging business with other countries.

Cumulative causation The process whereby a significant increase in economic growth can lead to even more growth as more money circulates in the economy.

In the model, economic growth begins in a region attractive to new industry.

+ Once growth begins in this 'core' region, flows of labour, capital and raw materials develop to support it and the growth region undergoes further expansion by the cumulative causation process.
+ A negative impact is felt in less developed regions (the periphery) as skilled labour and locally generated capital is attracted away.
+ Goods and services produced under the economies of scale of the core region flood the market of the periphery, undercutting local smaller scale enterprises.
+ The wealth gap between the core and the periphery widens (Figure 9.2.3).

However, increasing demand for raw materials from resource-rich parts of the periphery may stimulate growth in such regions. This may set off the process of cumulative causation, leading to the development of new centres of self-sustained economic growth. If the process is strong enough and significant economic growth occurs in the periphery, the inequality between core and periphery may begin to narrow.

Many developing countries are in the stage where the wealth gap between core and periphery is still widening. Thus, they have a high Gini coefficient.

Factors influencing inequalities within countries

+ **Residence**: regional differences; urban/rural disparities; intra-urban contrasts.
+ **Ethnicity**: the lower socio-economic status of many minority ethnic groups, often the result of discrimination.
+ **Employment**: jobs in the formal sector provide better pay and greater security than jobs in the informal sector.
+ **Education**: higher levels of education generally lead to better paid employment.
+ **Land ownership**: the distribution of land ownership (tenure) has had a major impact on disparities in many countries.

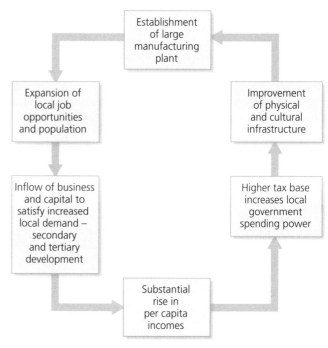

Figure 9.2.2 Simplified model of cumulative causation

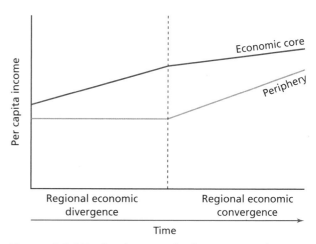

Figure 9.2.3 Regional economic divergence and convergence

> **Economies of scale** The cost reductions per unit of production achieved when companies increase the scale of production.

Exam tip

When studying world maps such as Figure 9.2.1, have an atlas at hand so that you can identify countries in each category of the legend (key).

Now test yourself

TESTED ○

1 Define neo-colonialism.
2 Describe social investment.
3 State three factors influencing inequalities within countries.

Impact of uneven development within a country: India

Regional contrasts

There is a wide variation in GDP per capita by state for 2017. The most advanced states/union territories (2017 data) include:

+ a group of states in the north-west, including Delhi (GDP per capita $5533), Chandigarh ($5088) and Haryana ($3433)
+ a line of states in the west and south, including Goa ($7045), Karnataka ($3139) and Kerala ($3097).

These two regions can be broadly viewed as the 'economic cores' of the country. There are lower levels of development in the northern and eastern states – the 'periphery'. The lowest GDP per capita states are Bihar ($640) and Uttar Pradesh ($950). States in the periphery often suffer from more difficult physical environments such as:

+ Rajasthan (far west, desert)
+ Uttar Pradesh (north, sub-Himalayan).

There is a major development gap in welfare and quality of life reflected in GDP and a range of other indicators:

+ **Poverty**: The World Bank in 2012 stated that a) the seven lowest-income states, with 45% of the population, housed 62% of India's poor; and b) 80% of India's poor lived in rural areas.
+ **Inadequate housing**: The average standard of housing in rural areas is well below that in urban areas and is often worse than in the large urban slums!
+ **Physical infrastructure**: Infrastructure indicators such as urbanisation, electrification and transport connectivity show huge differences between core and periphery regions in India.
+ **Unemployment**: Recent data shows a poor correlation between GDP per capita and unemployment. You might suggest why!
+ **Literacy**: In 2011, literacy in India was highest in Kerala at 94% and lowest in Bihar at 64%.
+ **Life expectancy at birth**: Ranges from 63 years in Assam to 75 years in Kerala.
+ **Other demographic rates**: The lowest rates of fertility and mortality are concentrated in the most economically advanced states.

Many socio-economic disparities are linked to a big gap in educational attainment and access to health services. Regional inequality has steadily come to the fore as a major political issue in India.

> **Life expentancy at birth**
> The average number of years a newborn infant can expect to live under current mortality levels.

> **Revision activity**
> To what extent do literacy and life expectancy at birth vary in India?

> **Exam tip**
> The correlation between indicators of development is rarely perfect. For example, a more affluent region may have relatively high unemployment due to a very high rate of in-migration from people seeking employment.

> **Now test yourself** TESTED
> 4 Where are India's two economic core regions?
> 5 Name India's two lowest GDP per capita states.
> 6 What physical factors have inhibited development in Rajasthan and Uttar Pradesh?

Uneven development and demographic data

The structure, or composition, of a population can be illustrated by the use of population pyramids which change in shape as countries progress through demographic transition and economic development (Figure 9.2.4). The four pyramids relate to the demographic transition model, with:

+ Niger in Stage 2
+ Bangladesh in Stage 3
+ the UK in Stage 4
+ Japan in Stage 5.

> **Population pyramid** A bar chart, arranged vertically, that shows the distribution of a population by age and gender.

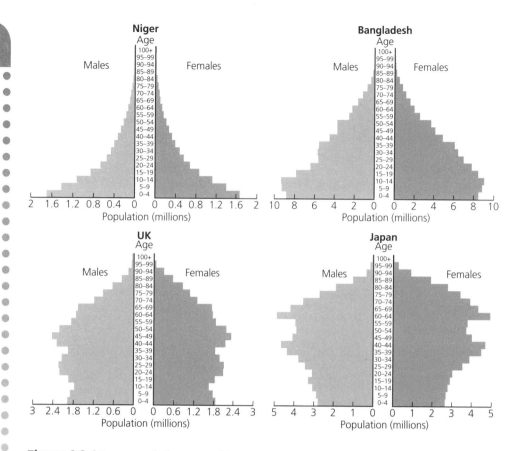

Figure 9.2.4 Four population pyramids

Figure 9.2.4 and Table 9.2.1 show that as development progresses:
+ The crude birth rate declines due to:
  + better education about, and improved access to, contraception
  + declining infant mortality due to advances in health care
  + increasing gender equality and employment opportunities for women
  + changing cultural expectations about family size.
+ The crude death rate falls initially, but then tends to rise slightly:
  + The death rate declines due to a range of economic and social improvements.
  + After a certain point, it tends to rise again because of the influence of an ageing population
+ The infant mortality rate falls sharply:
  + Age-specific mortality rates such as the infant mortality rate fall continually with development, particularly with investment in health, education and housing.
  + Such rates are not affected by the ageing of a population.
+ The maternal mortality rate declines:
  + Over 90% of all maternal deaths occur in low- and lower-middle income countries.
  + Skilled care before, during and after childbirth can save the lives of women and new-born infants.
+ Age structures change:
  + The young, dependent population (under 15 years) declines with development as fertility rates fall.
  + An increasing number of countries are below replacement level fertility.
  + The elderly dependent population increases as life expectancy improves.

Crude birth rate The number of live births per thousand population in a year.

Crude death rate The number of deaths per thousand population in a year.

Infant mortality rate The number of deaths of infants under 1 year of age per 1000 live births in a given year.

Maternal mortality rate The annual number of deaths of women from pregnancy-related causes per 100,000 live births.

Replacement level fertility A fertility rate of 2.1 children per woman is required to maintain a population at its current level. Below this level, a population will eventually decline.

Table 9.2.1 GNI per capita and demographic data for Niger, Bangladesh, the UK and Japan

Country	GNI per capita (US $)	Birth rate (per 1000)	Death rate (per 1000)	Infant mortality rate (per 1000)	Maternal mortality (per 100,000)	Population aged under 15 years (%)	Population aged over 65 years (%)
Niger	950	49	9	56	553	50	3
Bangladesh	3550	20	5	38	176	33	6
UK	40,550	12	9	4	9	18	27
Japan	38,870	8	10	2	5	13	27

Exam tip

When describing and explaining population pyramids, a good starting point is to divide the pyramid into three sections: the young, dependent population; the economically active population; the elderly, dependent population. You can then comment on each section in turn.

Revision activity

List four factors that contribute to the decline in the crude birth rate.

Now test yourself   TESTED ○

7 Calculate the rate of natural change for each of the four countries in Table 9.2.1.

8 Define the infant mortality rate.

9 Explain replacement level fertility.

9.3 Sustainable strategies to address uneven development

The range of international strategies   REVISED ○

International aid

International aid is assistance to countries in need in the form of grants or cheap loans. Most developing countries have been keen to accept foreign aid because many: a) lack the hard currency to pay for vital imports; b) have insufficient savings to invest in industry and infrastructure; and c) lack important technical skills.

The basic distinction in international aid is between official government aid and voluntary aid (Figure 9.3.1) run by non-governmental organisations (NGOs). The main concerns about aid are:
+ A significant proportion of aid is tied aid.
+ Too often aid fails to reach the very poorest people.
+ The use of aid on large, capital-intensive projects may worsen local poverty.
+ Aid may delay the introduction of reforms, e.g. the substitution of food aid for land reform.
+ Aid can create a culture of dependency.

International aid The giving of resources (money, food, goods, technology, etc.) by one country or organisation to another poorer country. The objective is to improve the economy and quality of life in the poorer country.

Non-governmental organisation An organisation (e.g. a charity) which is distinct from governmental or inter-governmental agencies.

Tied aid Tied aid is foreign aid that must be spent in the country providing the aid (the donor country).

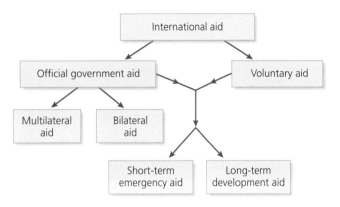

Figure 9.3.1 The different types of international aid

NGOs have often been much better at directing aid towards sustainable development than government agencies. The selective nature of such aid has targeted the poorest communities using intermediate technology and involving local people in decision making.

Intergovernmental agreements

The important roles of the WTO, the World Bank and the International Monetary Fund have already been covered. The WB and the IMF are both United Nations (UN) organisations. Other UN organisations also play a major role in development. These include the:

+ World Health Organization (WHO)
+ Food and Agriculture Organization (FAO)
+ United Nations Conference on Trade and Development (UNCTAD).

There is a strong relationship between trade and development. UNCTAD has stated that even small changes in agricultural employment opportunities, or farm prices, can have major socio-economic effects in developing countries.

> **Now test yourself**   TESTED ◯
>
> 1 Why are most developing countries keen to accept international aid?
> 2 State two possible disadvantages of international aid.
> 3 What is intermediate technology?

> **Intermediate technology**
> Aid supplied by a donor country whereby the level of technology and the skills required to service it are properly suited to the conditions in the receiving country.

> **Revision activity**
> Draw a diagram to show the different types of international aid.

> **Exam tip**
> UNCTAD is a very important UN organisation which aims to stimulate economic development through trade.

Different views on tackling the development gap

REVISED ◯

The development gap is such a major global issue it is not surprising that there are different views held by governments, organisations and individuals.

+ Some countries contribute a much higher proportion of their GDP to international development than others.
+ The political character of a government can influence its attitude to international development.
+ The strength of global civil society in a country can be a big factor in shaping development policy.
+ The respective benefits of prioritising trade or aid have long been debated.

> **Global civil society**
> Organisations or individuals, independent from the state, whose aim is to transform policies through communal efforts at a national or global scale.

Reforming the WTO

Critics of the WTO complain it is dominated by the largest economies in the world, with little regard to the trade difficulties faced by many developing countries. Giving developing countries greater access to markets in developed countries would be a major boost to the incomes of poorer countries.

Debt relief

+ Restructuring the debt of the poorest countries began in a limited way in the 1950s.
+ The heavily indebted poor countries (HIPC) initiative was established in 1996 by the IMF and World Bank.
+ In 2006, the multilateral debt relief initiative (MDRI) was launched to provide extra support to HIPCs to reach the millennium development goals.
+ Debt relief frees up resources for social spending.

Fair trade

Many large supermarkets and other large stores in developed countries now stock some 'fair trade' products. NGOs would like to see this system increased in scale.

Microcredit and social business

The development of the Grameen Bank in Bangladesh has illustrated the power of microcredit in the battle against poverty. The bank provides tiny loans and financial services to poor people to start their own businesses. Women are the beneficiaries of most of these loans. The concept has spread well beyond Bangladesh.

Muhammad Yunus has highlighted 'social business' as the next phase in the battle against poverty. He presents a vision of a new business model which combines the operation of the free market with the quest for a more humane world.

> **Exam tip**
>
> The aim of reforming organisations such as the WTO is to make them work better, particularly for their weaker members. However, it can be difficult to get all countries to agree to changes.

> **Now test yourself** TESTED ◯
>
> 4 What is global civil society?
> 5 Define debt relief.
> 6 How does microcredit work?

> **Debt relief** Cancellation of debts owed by developing nations to developed countries or institutions such as the World Bank, in order to allow the government to shift funds towards social development.
>
> **Fair trade** An institutional arrangement designed to help producers in developing countries achieve better trading conditions.
>
> **Microcredit** Tiny loans and financial services to help the poor, mostly women, start businesses and escape poverty.
>
> **Social business** Forms of business that seek to profit from investments that generate social improvements and serve a broader human development purpose.

> **Revision activity**
>
> Look on the internet to see which countries give the highest proportion of their GDP to international development.

Top-down and bottom-up development

Over the years much debate has focused on the amount of aid made available. More recently there has been greater discussion about the effectiveness of aid. Top-down approaches have been increasingly criticised. The Hunger Project, an NGO, is one of a number of organisations that have adopted a radically different approach. The Hunger Project has worked in partnership with local organisations in Africa, Asia and Latin America to develop effective bottom-up strategies. The key strands in these bottom-up strategies have been:

+ mobilising local people for self-reliant action
+ intervening for gender equality
+ strengthening local democracy.

Top-down development: The UK's High Speed 2 (HS2) project

HS2 is a planned high-speed railway linking London to cities in the Midlands and the North. The main objective of this massive project is to reduce regional economic imbalance. Cities in the North and Midlands have slower growth, lower productivity and lower wages than London and the South East. Poor transport infrastructure is a critical factor in this disparity. Apart from the high-speed service that HS2 will bring, it will also free up significant capacity on existing railway lines. HS2 is seen as a prerequisite to Northern Powerhouse Rail, a major West to East rail project. High-speed railways have been built in other European countries to reduce regional disparities.

+ Major construction began in 2018, but the majority of Phase 1 of the project remains to be completed. The first phase, linking London to Birmingham, should open in late 2026. The London terminal will be at Euston station.
+ Services on Phase 2A (to Crewe) are due to begin in 2027, with Phase 2B (to Leeds and Manchester) starting in 2033. Final decisions remain to be made about the route in certain places.
+ Trains will operate at speeds of up to 225 mph (362 km/h), reducing journey times considerably.
+ A recent estimate put the total cost at over £90 billion.

Critics of the project argue that HS2:
+ is far too expensive and that the money could be spent in a much better way
+ final cost estimates have risen sharply since the project was first proposed
+ will have an unacceptably large environmental impact
+ that the consultation process was largely cosmetic.

It would seem that public opinion is much more in favour of the project in the Midlands and North compared to the South. At the time of writing, the whole project is being reviewed by the government.

Bottom-up development: WaterAid in Mali

WaterAid has been active in the West African country of Mali since 2001. It is one of the world's poorest nations. In Mali:
+ over 4 million people still lack access to clean water
+ over 11 million people lack access to a decent toilet.

One of WaterAid's projects is in the slums surrounding Mali's capital Bamako, providing clean water and sanitation services to the poorest people.
+ WaterAid has financed the construction of the area's water network.
+ It has trained local people to manage and maintain the system, and to raise the money needed to keep it operational.
+ The community has been encouraged to invest in its own infrastructure, which is an important part of the philosophy of the project.
+ Significant improvements in the general health of the community have taken place.

WaterAid is dedicated to the provision of safe domestic water, sanitation and hygiene education to the world's poorest people. Without these three crucial elements, communities remain stuck in a cycle of disease and poverty (Figure 9.3.2). It takes a generation for health and sanitation to be properly embedded into people's day-to-day life.

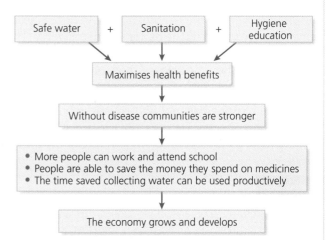

Figure 9.3.2 WaterAid's building blocks of development

In areas with WaterAid projects, life in times of drought is eased because:
+ Women, in particular, would spend hours in search of water, leaving little time to find food.
+ Children would miss out on education in the search for water.
+ Cattle can be watered, rather than sold or left to die because of water shortage.
+ During famines, with sanitation, water and hygiene, people are less sick and so are better able to fend off disease.

Now test yourself

TESTED

7 List the major cities in the UK to be connected by HS2.

8 Describe the geographical location of Mali.

9 What are WaterAid's primary objectives?

Exam practice

1 An improvement in life expectancy is part of which of the following categories of development? (1 mark)

   A economic

   B social

   C environmental

   D political

2 a) What is an emerging economy? (2 marks)

   b) Give two examples of emerging economies. (2 marks)

   c) Name the stages of development prior to, and after, emerging economy. (2 marks)

3 a) State two indicators of development used in the human development index (HDI). (2 marks)

   b) Name one country with very high human development. (1 mark)

   c) Name one country with low human development. (1 mark)

4 Explain the relationship between the stages of economic development and demographic data. (6 marks)

5 Examine the difference between top-down and bottom-up development. (6 marks)

6 Discuss the possible advantages and disadvantages of international aid. (12 marks)

Total: 35 marks

Summary

✚ Development, or improvement in the quality of life, is a wide-ranging concept.

✚ A range of factors contribute to the development and human welfare of a country.

✚ Food and water security are vital aspects of human welfare.

✚ Measures of development include GDP per capita, the human development index, and indices of political corruption.

✚ Historical, economic, social and other factors have resulted in uneven development between and within countries.

✚ The impact of uneven development is evident in poverty, unemployment, inadequate housing and poor physical infrastructure.

✚ There is a strong relationship between levels of economic development and demographic data.

✚ A range of international strategies including development aid have been used to try to reduce uneven development.

✚ Governments, organisations and individuals may have different views about how to tackle the development gap.

✚ There are advantages and disadvantages to both top-down and bottom-up approaches to development.

Index

Answers to Now Test Yourself and Exam Practice questions at **www.hoddereducation.co.uk/myrevisionnotesdownloa**

My Revision Notes: Pearson Edexcel International GCSE Geography

Notes